Amateur Antenna Tests and Measurements

by

Harry D. Hooton, W6TYH

Howard W. Sams & Co., Inc.
4300 WEST 62ND ST. INDIANAPOLIS, INDIANA 46268 USA

International Standard Book Number: 0-672-21466-0
Library of Congress Catalog Card Number: 77-083973

Printed in the United States of America.

Preface

The author has long felt the need for a complete reference manual on antenna measurements. Numerous articles on the subject have appeared in both amateur and professional technical journals, but these are not generally available to the average amateur. Much information with regard to antenna tests and measurements is passed back and forth on the ham bands, but the advanced experimenter often has difficulty in finding a fellow ham with the same interests and objectives.

This book is based largely on the author's own experiences over the years but it also contains many suggestions from other hams. The emphasis is on practical procedures rather than theory. The chapter on transmission-line fundamentals is very important since most modern antenna systems depend on line sections for phasing and power distribution. The success or failure of a complex array may depend upon your knowledge of transmission-line characteristics.

Every effort has been made to make the book readable and to present the essential information without difficult mathematical treatment. Wherever possible, the data is given in the form of tables and charts.

The test equipment and accessories described in this book are low-cost and easy to build. To enable you to obtain identical parts, a list of the component manufacturers and their mailing addresses is included in Appendix C.

HARRY D. HOOTON

Contents

CHAPTER 5

CHAPTER 6

CHAPTER 7

APPENDIX A

APPENDIX B

APPENDIX C

APPENDIX D

CHAPTER 1

Basic Antennas

The basic purpose of an antenna is to transform alternating current into an electromagnetic wave, or to transform electromagnetic wave energy back into an alternating current. The transmitting antenna produces electromagnetic waves; the receiving antenna collects them. In most two-way radio installations, the same antenna is used for both transmitting and receiving. The simple antenna is a special kind of resonant circuit that contains inherent inductance, capacitance, and resistance. The combination of these three properties produces another property called impedance. Antenna resistance contains both *ohmic* resistance, the kind that you can measure with an ohmmeter, and *radiation* resistance which must be measured with special test instruments designed for the purpose. In addition to these basic properties, the more elaborate and complex antenna arrays are *timed* or *phased* to produce effective power gain and directional characteristics. All of these antenna properties are measurable in practical units of ohms, volts, degrees, meters, feet, or inches. The purpose of this book is to tell you how to track down and measure these elusive and mysterious antenna properties. We will also show you how to build some simple and effective antenna test instruments to make the job easier. However, before we become involved in the measurement of antenna properties, we should take a good look at basic antenna and transmission line theory. Without a clear understanding of both, measurements on an actual antenna system will be meaningless.

1-1. ANTENNA FUNDAMENTALS

When a radio-frequency current is passed through a conductor, an alternating electromagnetic field is formed around it. This electromagnetic field consists of two components—a set of electric lines of force parallel with the conductor, and a set of magnetic lines of force that are perpendicular to the conductor and the electric lines of force.

If the conductor is a single straight wire one-half as long as the electromagnetic wavelength, it is called a *half-wave dipole antenna* (Fig. 1-1A). When the linear conductor is only one-

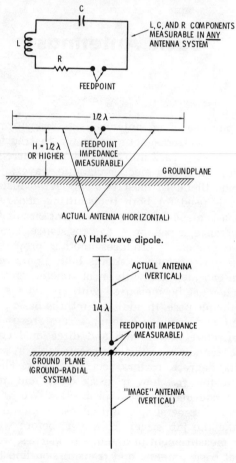

(A) Half-wave dipole.

(B) Quarter-wave monopole.

Fig. 1-1. Basic antenna systems.

fourth as long as the electromagnetic wavelength, it is called a *quarter-wave monopole antenna* (Fig. 1-1B). The actual antenna conductor (wire or rod) is often called an *element*. While dipole and monopole elements are often used singly as complete radiating systems, the modern practice tends toward the use of many elements, dipole or monopole, combined to form a more complex radiating system called an *array*. In order to obtain optimum performance from an array, certain measurements and adjustments of the electrical characteristics of the elements and of the phasing, power-divider, and transmission lines are necessary. To make these measurements and adjustments, especially those of a complex multielement array, you should have some experience in the use of basic test equipment such as a vtvm (or corresponding transistorized instrument), the oscilloscope, swr bridges, etc. If you are inexperienced in the use of these instruments, practice making measurements on a simple antenna such as a quarter-wave monopole before attempting to measure and adjust the more elaborate systems.

The basic antennas, when suspended in space or erected over a suitable ground plane, exhibit some of the properties of the ordinary resonant circuit. They contain inductance, capacitance, and *ohmic resistance* distributed over a wide area. In addition to the three properties just mentioned, there is one other property of the antenna that we must consider which is the *radiation resistance*. This is a theoretical resistance which would dissipate the same amount of power, in heat, as is radiated from the antenna in the form of electromagnetic energy. This *desirable* power loss due to radiation by the antenna is highest when the antenna conductor is resonant at the frequency of transmission. The greatest amount of rf energy from the transmitter will be radiated by the antenna when the transmission line impedance is *matched* to the radiation resistance, or to be more specific, to the impedance of the antenna.

The feedpoint impedance of a half-wave dipole antenna suspended in space, or erected horizontally one-half wavelength above a "perfect" ground, is theoretically 72 ohms (Figs. 1-2A and B). The base feedpoint impedance of a quarter-wave monopole antenna with an effective ground plane or radial system is approximately 36 ohms. When the half-wave dipole, resonant and operating under free-space conditions, is fed at the center with a 72-ohm transmission line, the antenna will appear as a *pure resistance* across the transmission line (Fig. 1-2C). Since the resistance (impedance) of the antenna feedpoint is equal to the *surge* impedance of the transmission line, the line and the antenna feedpoint are said to be matched in impedance. In a matched system, the rf current which passes along the line from transmitter to antenna "sees" the antenna as a fixed pure resistance, and all of the power that reaches the antenna end of the line is dissipated (most of it is radiated) by the antenna and no electrical energy is reflected back along the line toward the transmitter.

In order to determine whether or not the antenna and transmission are operating under matched conditions, you can measure

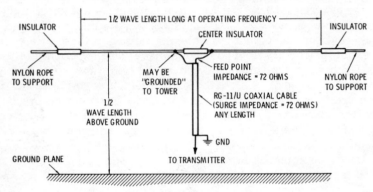

(A) A matched antenna system.

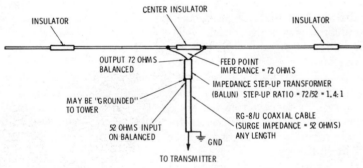

(B) A matched antenna system using an impedance-matching transformer.

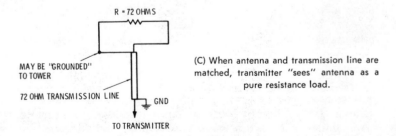

(C) When antenna and transmission line are matched, transmitter "sees" antenna as a pure resistance load.

Fig. 1-2. Matched Antennas.

the line swr (standing wave ratio) with a device called a directional coupler (Fig. 1-3) and a low-range dc voltmeter. Normally, using the directional coupler, the meter sensitivity is adjusted for a full-scale reading on the *forward* wave (going to the antenna) and then switched to read the *reverse* wave (going from the antenna back to the transmitter). Under matched conditions, the magnitude of the reverse wave will be zero and the line swr is 1:1. The directional coupler is described fully in a later chapter.

In the chapters that follow, we will discuss simple test equipment for measuring impedance, phase, gain, and other antenna-

Fig. 1-3. Author using a directional coupler to tune-up a directional array.

system characteristics. The test and measurement procedures are relatively simple and easy to perform. They require no high degree of technical education or skill on the part of the radio amateur.

A set of ten "test" questions, with true and false answers, is included at the end of each chapter. Use the questions to test your knowledge of the subject under discussion.

CHAPTER 2

Radio-Frequency
Transmission Lines

Most radio amateurs are familiar with the use of radio-frequency transmission lines to transfer energy from one point to another, usually from an antenna system to a receiver, or from a transmitter to an antenna system. Some amateurs are also familiar with the use of transmission lines as circuit elements such as impedance transformers, phase inverters, wave filters, tank circuits, etc. It should also be realized that *anything* that is employed to transfer radio-frequency energy from one point in a system to another point in the same system will obey certain fixed laws and will possess certain basic characteristics that will affect the overall performance of the system. In a practical system, such as a multielement phased antenna array, it is virtually impossible to predict with any degree of accuracy the exact conditions that will be found at different points in the system when it is completed. The usual procedure is to design and construct the system as closely as possible to the theoretical conception. Then, by means of extensive measurements on the completed system, determine what corrections, changes, or modifications will be necessary in order to achieve the original performance goals. A good understanding of the basic characteristics of transmission lines is essential for an intelligent measurement and adjustment of antenna-system behavior.

2-1. GENERAL BEHAVIOR OF TRANSMISSION LINES

While transmission lines have various configurations, such as parallel wire, coaxial cable, balanced lines, and unbalanced lines, their general behavior is the same. Most amateur antenna systems are fed radio-frequency energy from the transmitter through co-axial cable, usually with a *surge impedance* of 52 or 75 ohms. What does the term surge impedance mean, precisely? Let us suppose that we have an idealized coaxial line consisting of inner and outer conductors having zero resistance, uniform cross section and spacing, and infinite length. Let us connect a switch and battery across one end of our infinitely long line as shown in Fig. 2-1. For purposes of simplification, let us assume that the internal resistance of the battery is zero.

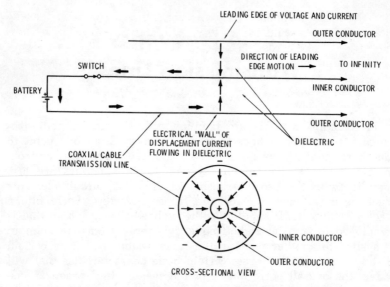

Fig. 2-1. Displacement current in an infinitely long transmission line.

When the switch is closed, the line will progressively charge along its length to the potential of the battery. The line *does not* instantaneously become charged throughout its entire length at the instant the switch is closed. The electrical action along the line is somewhat analogous to the action of a dam suddenly break-line causing a "wall" of water to flow down a valley. The electrical "wall" in the line is the boundary or point of division between the charged and uncharged portions of the line. The dividing point starts at the exact instant when the switch is closed and travels down the line at a rate equal to the characteristic velocity of space times the velocity constant of the line.

The characteristic velocity of a radio wave traveling through free space is about 186,000 miles per second. The velocity constant of coaxial cables commonly used by radio amateurs (RG-8/U and

RG-11/U, for example) is 0.66. This means that a radio wave will travel only 66 percent as fast inside the coaxial cable as it does in air or free space. The electrical "wall" or point of division actually represents the point at which a *displacement current* flows in the space between the two conductors during any given instant. In order to maintain this displacement current, a steady *conduction current* from the battery flows down one conductor of the line, through the boundary or point of division, and back to the battery through the other conductor. The current and voltage at the boundary point, traveling down the line at a velocity somewhat less than the speed of light, is most generally referred to as a *leading edge* of a radio wave. In the section of line between the leading edge and the battery, the flow of conduction current and the voltage across the line are uniform in value. In the section of cable ahead of the leading edge, the voltage and current values are zero. The important thing to remember is that it takes *time* for the leading edge of a radio wave to travel in air, space, or in a transmission line.

A phased antenna system is simply a system in which the instant of radiation of some portion of the radio wave is delayed at one or more elements of the system as compared to the instant of radiation at some other element. A phased radiating system causes the antenna elements to "fire" at different time intervals with respect to each other in a manner somewhat similar to the firing order of a multicylinder gasoline engine. Just as the gasoline engine must be correctly timed to obtain maximum performance and efficiency, the multielement antenna array must be correctly timed or phased if optimum performance and efficiency are to be realized.

Let us continue our discussion of the infinitely long transmission line. In order to charge a 122,760-mile section of RG-8/U coaxial cable in a time of one second, the battery must supply a definite amount of current. The current flow from the battery will begin at the instant the switch is closed and will remain constant as long as the switch is closed. The displacement current at the leading edge is constant in value at all times even though it is in motion down the line. It is apparent that, to the battery, the infinitely long, uniform, lossless transmission line "looks" like an impedance with a net reactance of zero, or a *resistance*.

We know that a transmission line, for purposes of analysis and explanation, is considered to be a special arrangement of lumped constants consisting of resistive, inductive, and capacitive elements. In the transmission line, however, the reactive elements are uniformly distributed throughout its entire length and the

X_C and X_L components are always such that they cancel each other. Since the reactive components are uniformly distributed, the only effect on the battery of changing the ratio of the inductance to capacitance, or vice versa, for a given length of time would be to change the amount of the current flow for a given applied voltage. In a practical coaxial line, the characteristic or surge impedance is generally specified in terms of the physical characteristics of the line shown in Fig. 2-2. The characteristic impedance is given by the equation:

$$Z_o = \frac{138}{\sqrt{k}} \log \frac{D}{d}$$

where,
   k is the dielectric constant,
   D is the inside diameter of the outer conductor,
   d is the outside diameter of the inner conductor expressed in the same units as D (inches or centimeters).

COPPER-BRAID OUTER CONDUCTOR

STRANDED WIRES
INNER CONDUCTOR

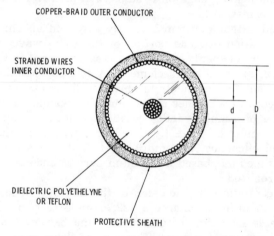

d   D

DIELECTRIC POLYETHELYNE
OR TEFLON

PROTECTIVE SHEATH

Fig. 2-2. Cross section of solid-dielectric coaxial line.

It is obvious that if the outside diameter of the inner conductor or the dielectric constant of the insulating material is increased, the capacitive component of the line will increase and the charging current drawn from the battery at a given potential value will increase. According to Ohm's law, when a voltage remains constant but the current increases through a circuit, the resistance must be lower. Therefore, to design a transmission line for a specified characteristic impedance (Z_o) both the ratio of the conductor diameters and the dielectric constant (K) must be closely controlled. The dielectric constant for air is 1 and for polyethylene is 2.25. The graph in Fig. 2-3 shows the characteristic impedance of air-dielectric coaxial lines with various conductor-diameter ratios. The reader should become familiar with this graph as it will be helpful in the design of the signal-sampler and directional-coupler devices described later in this book. It is suggested that the reader carefully examine the cross-sectional area of standard RG-8/U (50 ohms) and RG-11/U (75 ohms) coaxial cables, noting that the inner conductor of the lower-impedance cable (RG-8/U) is larger in diameter than that of the 75-ohm RG-11/U.

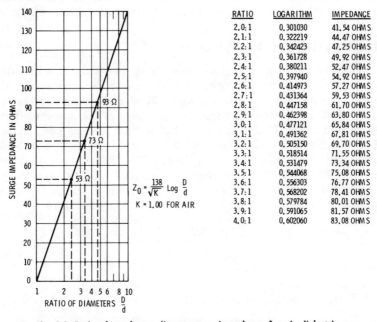

RATIO	LOGARITHM	IMPEDANCE
2.0:1	0.301030	41.54 OHMS
2.1:1	0.322219	44.47 OHMS
2.2:1	0.342423	47.25 OHMS
2.3:1	0.361728	49.92 OHMS
2.4:1	0.380211	52.47 OHMS
2.5:1	0.397940	54.92 OHMS
2.6:1	0.414973	57.27 OHMS
2.7:1	0.431364	59.53 OHMS
2.8:1	0.447158	61.70 OHMS
2.9:1	0.462398	63.80 OHMS
3.0:1	0.477121	65.84 OHMS
3.1:1	0.491362	67.81 OHMS
3.2:1	0.505150	69.70 OHMS
3.3:1	0.518514	71.55 OHMS
3.4:1	0.531479	73.34 OHMS
3.5:1	0.544068	75.08 OHMS
3.6:1	0.556303	76.77 OHMS
3.7:1	0.568202	78.41 OHMS
3.8:1	0.579784	80.01 OHMS
3.9:1	0.591065	81.57 OHMS
4.0:1	0.602060	83.08 OHMS

$$Z_0 = \frac{138}{\sqrt{K}} \ \text{Log} \ \frac{D}{d}$$

$K = 1.00$ FOR AIR

Fig. 2-3. Ratio of conductor diameters vs. impedance for air dielectric.

The characteristic impedance of a specific segment of a line is the same as the impedance of the entire line. In other words, 100 feet of RG-8/U (50 ohms) *does not* have twice as much impedance as a 50-foot segment of the same line. However, for some applications, such as phasing, line length is a very important factor.

In practical application, infinitely long transmission lines do not exist in the physical sense. In the electrical sense, however, the characteristic equivalent of an infinitely long line is produced by terminating a transmission line in a *resistance* equal in value to the surge impedance of the line. This resistance may or may not be an actual ohmic resistance. In most cases, the terminating resistance may be an antenna or a transformer adjusted so that, when connected across the line, the line "sees" it as a resistance. When an antenna radiates substantially all the radio-frequency power that the transmitter supplies to the transmission line, and no power is reflected back to the source, the system is said to be *matched*. In a matched system, the line power at the terminating point is completely absorbed by the antenna. In a receiving system, the antenna is the source of energy and the line is considered terminated at the receiver input terminals. For transmitting, the line is actually terminated at the antenna feedpoint through an adjustable impedance-matching network to be discussed later.

2-2. WAVELENGTH AND VELOCITY

When working with transmission lines, particularly when designing phasing-line networks or making swr measurements, always consider the *velocity constant*. Most radio amateurs know that a radio wave travels through free space at the speed of light —approximately 186,000 miles, or 300 million meters, per second. The velocity of a radio wave in air is, for all practical purposes, the same as its velocity in free space. The velocity constant for air is 1.00. In all other dielectric materials, radio waves travel slower than in free space or air. Mathematically, the radio-wave velocity of propagation in *any* medium is approximately:

$$V = \frac{2.998 \times 10^{10}}{\sqrt{k}}$$

where,
   V is the wave velocity in centimeters per second,
   k is the dielectric constant of the medium.

or,

$$V = \frac{299,800,000}{\sqrt{k}}$$

where,
   V is the velocity in meters per second.

Insulating materials used in transmission lines, particularly co-axial lines used by radio amateurs, have dielectric constants greater than 1.00. Therefore, the propagation velocity of a radio wave along any transmission line is always less than the wave velocity in free space or air. Table 2-1 shows the characteristics of various insulating materials commonly used by radio amateurs.

Table 2-1. Dielectric Constants of Common Insulating Materials

Material	1 MHz	100 MHz
Porcelain (dry process)	5.08	5.04
Fused quartz	3.78	3.78
Steatite 410	5.77	5.77
Foamed polystyrene	1.03	1.03
Phenol (Bakelite BM120)	4.36	3.95
Polyethylene	2.26	2.26
Polypropylene	2.55	2.55
Polyvinyl-chloride (PVC)	2.88	2.88
Transil oil	2.22	2.20
Neoprene rubber	6.26	4.50
Ruby mica	5.40	5.40
Soil, sandy dry	2.59	2.55
Soil, loamy dry	2.53	2.48
Distilled water	78.20	78.00

Table 2-2 gives the characteristics of the most commonly used commercial coaxial and parallel-conductor transmission lines.

Radio waves are produced by the radiation of high-frequency, alternating-current energy. Since the waves travel in a given direction at a constant velocity (depending upon the medium), they *must* cover a certain linear distance during the time period of one cycle. The distance the wave travels during the one-cycle time period is called the *wavelength*. The velocity of the wave equals

Table 2-2. Characteristics of the Most Commonly Used Coaxial and Parallel-Conductor Transmission Lines

Type	Impedance in Ohms (nominal)	Dielectric Type	VC	Maximum RMS Amperes
RG-8/U C*	52	A†	0.66	7 to 10
RG-58/U C	53	A	0.66	2.8 to 5
RG-11/U C	75	A	0.66	5 to 7
RG-59/U C	75	A	0.66	2.8 to 5
RG-17/U C	52	A	0.66	12 to 15
RG-62/U C	93	A	0.84	— — —
RG-71B/U C	93	A	0.84	— — —
TWIN LEAD S (TV-FM)	300	A or plastic ribbon	0.82	— — —
No. 12 0W6	600	Air	0.975‡	— — —
No. 14 0W6	550	Air	0.975	— — —
No. 14 0W4	500	Air	0.975	— — —

* C indicates coaxial line
† A indicates polyethylene
‡ Impedance and VC (velocity constant) calculated and estimated for use with porcelain spreader insulators spaced at 4-foot intervals along line
0W4 and 0W6 indicate "open-wire 4-inch spaced" and "open-wire 6-inch spaced" lines, respectively

the distance covered during one cycle (wavelength) divided by the time of *one* cycle. Therefore:

$$f = \frac{1}{T}$$

where,

f is the frequency in hertz,
T is the time in seconds.

and,

$$V = \lambda f$$

where,

V is the velocity in meters or feet per second,
λ is the wavelength in meters or feet,
f is the frequency in hertz.

From these basic equations, other useful expressions are derived. These expressions will be covered later.

When discussing the electrical behavior of transmission lines, it is more convenient to speak in terms of *electrical length* rather than physical length. By this time, the reader should have firmly established in his mind that the velocity of a radio wave is determined *only by the properties of the medium through which it is passing* and is in no way affected by its frequency or wavelength. As pointed out above, when the transmission-line insulating materials (medium) have a dielectric constant greater than 1.00, the propagation velocity of the wave is reduced.

To determine the *physical length* of a line, multiply the free-space length by the velocity constant. To determine the free-space wavelength in *meters:*

$$\lambda = \frac{300}{f\,(\text{MHz})}$$

where,

λ is the wavelength in meters,
f is the frequency in megahertz.

To determine the free-space wavelength in *feet:*

$$\lambda = \frac{984}{f\,(\text{MHz})}$$

where,

λ is the wavelength in feet,
f is the frequency in megahertz.

In order to reduce mathematical expressions and calculations to the minimum, we have compiled the data given in Table 2-3.

Table 2-3. Free-Space (Air Dielectric) Wavelengths in Feet for Amateur-Band Center Frequencies, 3 to 30 MHz

Degrees	3.75 MHz	7.15 MHz	14.175 MHz	21.225 MHz	28.6 MHz
360	262.39	137.62	69.42	46.28	34.40
270	196.80	103.21	52.06	34.80	25.80
180	131.20	68.82	34.70	23.18	17.20
90	65.60	34.41	17.35	11.60	8.61
45	32.80	17.21	8.68	5.80	4.30
22.5	16.40	8.60	4.35	2.91	2.15
10.0	7.28	3.82	1.92	1.29	0.95
5.0	3.65	1.91	0.97	0.65	0.48

$$\text{Wavelength (feet)} = \frac{984}{\text{Frequency (MHz)}}$$

See Appendix for additional information

The wavelengths given for key frequencies in the 10-, 15-, 20-, 40-, and 80-meter amateur bands are free-space or air-dielectric wavelengths. To obtain the actual physical wavelength of a given line, simply multiply the given free-space wavelength by the velocity constant of the line to be used. For example, suppose that we wish to determine the actual physical length in feet required for one electrical wavelength of RG-8/U coaxial cable at 7.2 MHz. From the equation for wavelength, we find that the free-space wavelength is 136.67 feet. One electrical wavelength of RG-8/U cable with standard polyethylene dielectric (not foam) will be equal to 136.67 feet multiplied by 0.66 (the velocity constant of polyethylene) or 90.2 feet. Simplified formulas for calculating different lengths of coaxial cables will be given later.

In our preliminary discussion, where we used a battery and switch to charge the infinitely long transmission line, it was pointed out that the leading edge of the voltage and current traveled down the line, away from the source, at some velocity less than the speed of light. In actual practice, coaxial transmission lines are not ordinarily used to transmit direct current. The energy transmitted along the line usually consists of high-frequency alternating current. Nevertheless, the basic principle remains. Since a wave traveling down a line has a constant velocity, it requires increasing amounts of time to reach successive points at greater distances from the source. The leading edge of a radio wave traveling down a line is generally referred to as a "wavefront." Radio-frequency current traveling over one electrical wavelength of line completes one full cycle of frequency. If the transmission line has been properly terminated, the phase relationship of the radio-frequency currents at various points along the line (compared with the source-current phase) is as shown in Fig. 2-4. Note that when the line current has traveled an elec-

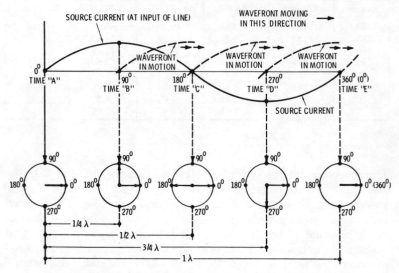

Fig. 2-4. Impulse wavefront moving down a transmission line.

trical quarter-wavelength down the line, the source current has completed 90° of the *next* cycle; when the line current wavefront arrives at the half-wavelength point, the source current has completed 180° of the next cycle; and when the line current has traveled one full wavelength down the line, the source current has completed the next cycle. The line current at a point one full wavelength down the line and the source current will go through their zero and maximum value points simultaneously in perfect synchronism and are said to be *in phase* with each other. The line current, however, is *delayed* by an amount of time equal to 1/f, or one complete cycle. In many amateur applications, this delay has no practical significance since both currents are sine waves. However, this delay principle must be thoroughly understood because it is the secret ingredient in the success or failure of phased antenna systems. Later, we will show you how to *measure* the radiation delay between various elements in a phased antenna system accurately.

2-3. PHASE SHIFT IN TRANSMISSION LINES

Since radio-frequency current travels at a constant velocity along the transmission line, a line one electrical wavelength long may be divided into electrical degrees, as shown in Fig. 2-5. In the previous example, we found that one electrical wavelength of RG-8/U coaxial cable is 90.2 feet long at 7.2 MHz. If the cable characteristics are uniform along its length (no sharp bends or dents) and it has been terminated in a pure resistance equal to its surge impedance, the electrical one-wavelength line may be divided into electrical degrees, as shown in Fig. 2-5. Since our

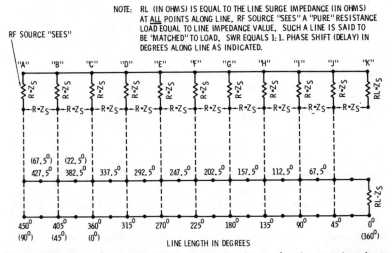

Fig. 2-5. Transmission line terminated in a pure resistance equal to its surge impedance.

90.2-foot line equals one electrical wavelength at 7.2 MHz, it also equals 360° at 7.2 MHz. If we divide 360° by 90.2 feet, we find that an RG-8/U coaxial cable at 7.2 MHz has a signal delay, or *phase shift*, of approximately 4° per foot. At this frequency, any section of RG-8/U or RG-11/U cable with standard dielectric "A" (solid polyethylene, not foam) will delay the radio-frequency current traveling down it at a constant rate of 4° per foot. This is a constant characteristic of the line at this frequency (7.2 MHz) and has nothing to do with the electrical or physical lengths of the line.

When using coaxial-cable or other transmission-line sections as delay networks, it is customary to refer to the delay as a *phase shift* and specify the phase shift in terms of *degrees*. When lumped-constant (fixed capacitance and inductance values) sections are used as delay networks, the delay may be specified in terms of time (microseconds, usually) instead of degrees. In either case, the electrical action occurring in the circuit is the same.

For the one-wavelength RG-8/U line at 7.2 MHz, the physical length of the line is 90.2 feet. Thus, the electrical half-wavelength point (180°) will be at the center of the line, or 45.1 feet down the line from the source. If the line signal at this point is compared with the source signal, it will be found that the two signals are exactly alike (both sine waves) but are precisely opposite in polarity (out of phase by 180°) at any given instant. The line signal at the electrical half-wavelength point is running exactly one-half cycle *late* with respect to the source signal. Any signal occurring at a time later than the reference source signal is said to be *lagging* the reference signal. Therefore, we may say that the signal located 45.1 feet down an RG-8/U or RG-11/U line from the source, when the source signal frequency is 7.2 MHz, is lagging the source signal by 180°. If the source frequency is changed, the 180° point along the line will also change.

It must *not* be supposed that RG-8/U or RG-11/U line has an inherent phase shift of 4° per foot; it has this value only at the frequency of 7.2 MHz. In a similar manner, the electrical quarter-wavelength point (90°) at 7.2 MHz will be found at 22.55 feet down the line from the source; and the three-quarter-wavelength point (270°) will be at 67.65 feet down the line from the source. Any phase-shift point (in degrees) may be found on the line by dividing the desired phase shift in degrees by the number of degrees per foot of line. As an example, suppose that we wish to delay a 7.2-MHz signal by 135° with respect to the source signal. Divide 135° (the desired delay) by 4° (the phase shift per foot) and the phase-shift point will be 33.75 feet down the line from the source.

By now, it should be obvious that the amount of phase shift introduced by the line is a function of the propagation velocity of the wave and the distance of the reference point from the wave source. When the propagation velocity of the wave is constant, the actual specified phase-shift point (degrees) in terms of linear measurement (meters, feet, etc.) will be a function of the operating frequency.

2-4. ODD AND EVEN MULTIPLES OF HALF-WAVELENGTHS

The current flowing down a transmission line containing a number of electrical full-wavelength sections will be in phase with the source current at all points down the line which are at consecutive electrical full-wavelength distances from the source. This phenomenon occurs because the line-current phase shift is 360°, or one cycle, with respect to the source current over *each* full-wavelength line segment. For example, the total phase shift over a line two wavelengths long is 720°, or two complete cycles of source current. It must be kept in mind, however, that although the line and source currents are in phase at the electrical two-wavelength point, the line current is actually delayed two complete cycles behind the source current.

In the past, it has been customary to speak of transmission-line lengths as being in multiples of half, rather than whole, wavelengths. Because most directional antenna systems were operated with the elements either *in phase* (0°) or *out of phase* (180°), the terms *odd* and *even* multiples of half-wavelengths came into widespread use. These terms are generally found in some of the antenna engineering books and reference manuals published many years ago. Today, most writers specify the phase shift in terms of degrees. However, it is interesting to note that the current in a transmission line equal to an odd multiple of an electrical half-wavelength (1/2, 3/2, 5/2, etc.) is 180° out of phase with the source current at the *odd half-wave* points; and the current of a line equal to an even multiple (2/2, 4/2, 6/2, etc.) is in phase (0° phase difference) with the source current at the *even half-wave* points.

2-5. LINE TERMINATION

Let us assume that a generator (transmitter) is "looking" into an infinitely long line with a characteristic, or surge, impedance of 50 ohms. The term *infinitely long* means that a radio wave traveling down this line goes on forever and is never reflected or returned back toward the source (generator). Now, if we cut or break this line at a distance of, say, 300 feet from the source and connect a 50-ohm pure resistance across the end of the line section still connected to the source, the source (transmitter) will "see" no difference between the *terminated* line and the original infinitely long line. Thus, a line that is infinitely long electrically may be relatively short physically when it is terminated in a *load resistance* that is equal to the characteristic, or surge, impedance of the line.

As mentioned previously, load resistance is not an actual resistor (except for testing purposes), but the radiation-resistance feedpoint impedance of the antenna conductor. However, it makes no difference what the load resistance actually is, as long as the transmission line "sees" a load which acts as a resistance equal to the surge impedance of the line. If the load is not a pure resistance equal to the surge impedance of the line, then it contains a *reactive component* (either inductive or capacitive) and reflections will occur at the load end of the line. Reflections will also occur if the load is a pure resistance but is *not* equal to the characteristic impedance of the line.

Reflected power is simply power that has been supplied to the line by the transmitter but has not been completely absorbed by the load. The power not absorbed by the load (antenna) is reflected back along the line toward the transmitter and creates standing waves of radio-frequency voltage and current along the line. The magnitude of the standing waves on the line depends upon the amount of mismatch between the characteristic impedance of the transmission line and the load impedance. While there is general agreement among amateurs that transmission lines should be matched to the load, there is often a great deal of confusion with regard to the reasons for matching and the methods employed. If intelligent antenna measurements and adjustments are to be made, the radio amateur must have a thorough understanding of standing wave formation and the meaning of the term *swr*.

The swr (standing-wave ratio) expresses the degree of mismatch existing between the line and its terminating load. The swr equals the ratio of maximum to minimum radio-frequency voltage or current along the line. When a transmission line is terminated in a resistance load equal to its characteristic impedance, no standing waves are exhibited, the voltage and current values along the line will be uniform, and the swr is 1:1. On the other hand, power applied to a transmission line terminated in either an open circuit (Fig. 2-6) or a short circuit (Fig. 2-7) results in a theoretically infinite swr, assuming that there is no lead resistance.

Let us see what happens when radio-frequency power is applied to a line with an open circuit. The voltage and current waves from the transmitter travel down the line until they arrive at the open-circuited output end. The impedance of an open circuit is considered to be practically infinite and since electrons cannot flow across a point of infinite impedance, the current wave must collapse. An electron current, no matter whether rf, ac, or dc, always creates a magnetic field around the conductor in which it flows. When the current wave collapses at the line open-circuit termination, the accompanying magnetic field also collapses. The collapsing magnetic field cutting across the line conductors induces a reverse voltage across the line. This starts voltage and current waves traveling back along the transmission line in the direction of the transmitter. These *reflected* waves traveling back along the line meet the waves traveling from the transmitter. Since the reflected wave is traveling back on the line at exactly the same velocity as the wave from the transmitter is traveling forward, the resultant wave does not move along the line, and is, therefore, called a standing wave.

The resultant voltage (or current) at any point on the line is equal to the sum of the voltage (or current) of the original wave from the transmitter at that point plus the voltage (or current) of the reflected wave at that point. The values of the forward wave (from the transmitter toward the load) or the reflected wave (from the load toward the transmitter) may be measured

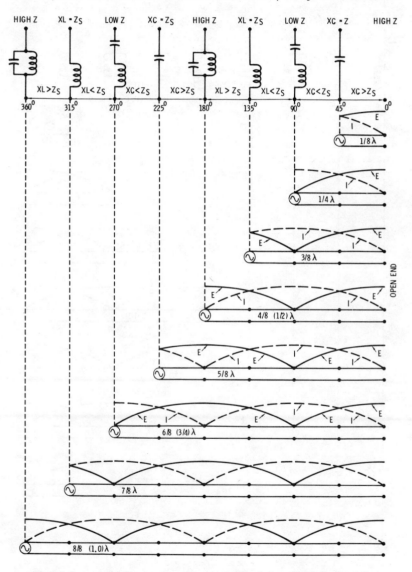

Fig. 2-6. Transmission line terminated in an open-circuit impedance characteristic.

at any point in a transmission line by opening the line and inserting a *directional coupler* into the line. The directional coupler is used with diode rectifiers and a dc indicating instrument, such as a low-range dc microammeter. The complete instrument is generally called a *monimatch,* a *reflectometer,* or possibly an *swr meter.* The uses of this instrument and its limitations in proper

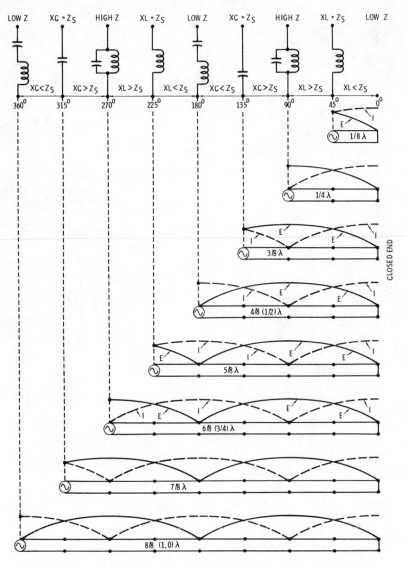

Fig. 2-7. Transmission line terminated in a short-circuit impedance characteristic.

antenna adjustment procedures will be discussed in the next chapter. Standing waves of *current* on a transmission line shorted at the far end are illustrated in Fig. 2-8.

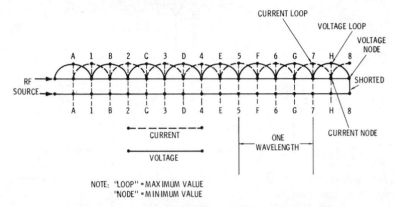

Fig. 2-8. Voltage and current nodes on a shorted transmission line.

2-6. STANDING-WAVE DISTRIBUTION ON TRANSMISSION LINES

If we assume that the line loss is zero, the current at the nodes will be zero as shown in Fig. 2-8. The current between the nodes varies throughout a cycle. The *instantaneous* value will vary from a maximum negative current value, through zero, and then up to a maximum positive current value during the time of each rf cycle. Fig. 2-9 illustrates the *effective values* of the current and voltage standing waves on a transmission line terminated in an open circuit at the load end of the line. Effective voltage and current values are always considered to be positive and are, therefore, shown above the horizontal reference line.

The voltage and current nodes in Fig. 2-8 are separated by 90°, or one-quarter wavelength. Notice that a current loop (maximum) is occurring at the short-circuited point while a voltage node (minimum) occurs at the same point. When a transmission line is terminated in an open circuit, exactly opposite conditions will exist. The voltage and current nodes will still be separated by 90° but will be displaced as shown in Fig. 2-9, with a current node and a voltage loop occurring simultaneously at the open end of the line. A short-circuit or open termination is considered to be the greatest possible mismatch between the line and load. The resulting standing waves are the highest that can be produced. However, the reader must not jump to conclusions that the power supplied to a line with a high swr is lost. At the lower frequencies, the current at the nodes will be very low; in a com-

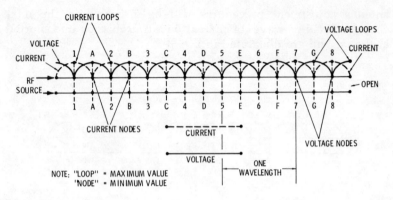

Fig. 2-9. Voltage and current nodes on an open transmission line.

pletely lossless line, the rf current will drop to zero at the nodes. No practical transmission line is completely without loss. However, the loss in power from an swr of 2:1, or even 3:1, is negligible in the frequency region from 3.0 to 30 MHz. The power loss from a high swr is greatest in the vhf and uhf bands where the increased rf resistance in the line due to skin effect becomes a major factor. At these frequencies, the highest swr that can be tolerated is about 2:1. In most cases, an swr of 3:1 and higher will result in excessive power loss at vhf and uhf.

2-7. SWR MEASUREMENTS

In the early days when open-wire transmission lines were in common use, an "swr meter" usually consisted of nothing more than a low-wattage neon lamp. When the glass envelope was brought in contact with one wire of the line, the lamp would glow, even with relatively low power. Today, most radio amateurs use coaxial transmission lines and, since the rf fields are confined within the cable, more sophisticated test instruments are required. In coaxial-cable lines, swr and forward and reflected power measurements are usually made by the insertion of a *directional-coupler* or *signal-sampler* section in series with the coaxial line at the point where the measurements are to be made. The use of the directional coupler is probably best illustrated by a brief discussion of an instrument in common use among amateurs, the *monimatch*.

The schematic diagram for a monimatch, which is often constructed as part of a variable impedance-coupling or matching unit, is shown in Fig. 2-10. As indicated in the schematic, the directional coupler consists of a flexible coaxial pickup line made from a 14-inch length of RG-8A/U. The outer vinyl cover is removed and a small hole is made in the outer braid 4 inches each side of center. The braided outer shield is then bunched up so that a No. 22 insulated wire may be inserted in one hole, pushed under the braid to the second hole, and then brought out the second hole, as shown in Fig. 2-10. Be careful not to cut or damage the small wire strands of the braided outer conductor. The braid is now smoothed out and covered with vinyl electrical tape. Check the No. 22 insulated wire to make certain that it is not shorted to the braided outer conductor of the coaxial cable. The two ends of the insulated pickup wire are connected to a ceramic

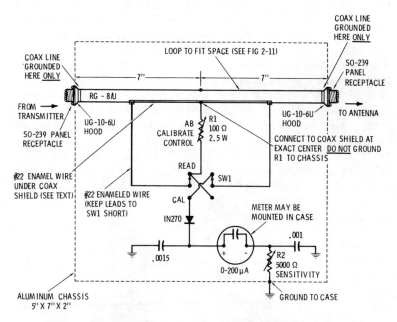

Fig. 2-10. Schematic diagram for the monimatch directional coupler.

rotary dpdt switch as shown in the schematic. The 100-ohm calibration potentiometer should be insulated from the case and mounted in such a manner that its capacitance to ground is reduced as much as possible. The indicating device is a 0–200-μA dc microammeter which may be included in the assembly or, if desired, connected externally. The ground connection for the 100-ohm potentiometer is returned to the center of the coaxial-cable outer shield. All other ground-return connections are made to the instrument case. The completed instrument is shown in Fig. 2-11.

To calibrate the instrument for use, connect a 52-ohm dummy load to the output terminal and feed an unmodulated rf carrier signal to the input terminal. With the switch in the CALIBRATE position, adjust the sensitivity control until the meter reads about half-scale. Switch to the READ position and adjust the sensitivity control for full-scale indication. The 100-ohm calibrate potentiometer is now adjusted for a null (minimum) indication. Alternately switch from CALIBRATE position to READ position, readjust the sensitivity control for full-scale meter reading, and adjust the 100-ohm calibrate potentiometer for a null. With a 52-ohm resistance dummy load, the null should be very close to zero. If the antenna is matched to the 52-ohm transmission line from the transmitter, the indicating meter should read approximately zero when the switch is placed in the READ or REVERSE position. Before taking a reverse reading, load the transmitter and adjust the sensitivity control for exactly full-scale indication on the meter. Now, switch to the READ or REVERSE position and note the meter reading. If the swr on the line is 2:1 or less, a very low meter reading will be observed. If the line is perfectly matched to the antenna, the meter should read zero when switched to the READ position. For

Fig. 2-11. Internal view of the monimatch directional coupler.

further details on the practical construction of a monimatch unit, refer to the *Radio Handbook* published by Howard W. Sams & Co., Inc.

Although some radio amateurs use the monimatch for making adjustments on antenna systems, it is most useful as an swr monitor and an indicating device for adjustment of an antenna coupler or matching unit. It has the disadvantage of being more sensitive on the higher-frequency bands and will provide full-scale meter readings only on relatively high-power systems, particularly when used on the 40- and 80-meter amateur bands. The test instruments to be described later, beginning with the next chapter, have been designed for use with very low-level source signals. One of these instruments, the *signal-sampler* section, functions in a man-

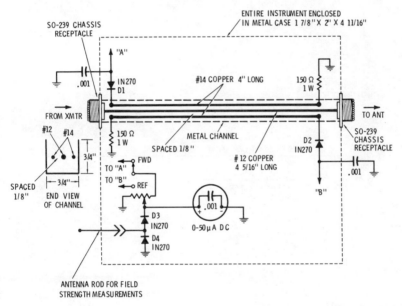

Fig. 2-12. Small swr meter for CB.

ner somewhat similar to that of the monimatch unit. However, the indicator is a vtvm with a diode detector probe, the lowest scale of which is 0-0.5 volts dc. The schematic diagram of a small imported swr meter for CB use is shown in Fig. 2-12. This instrument is useful for measuring swr in amateur antennas. However, it is designed for very low-power operation (about 3 or 4 watts) and is easily damaged. Even though these instruments are quite inexpensive, they are sufficiently accurate for low-power amateur radio swr measurements.

2-8. HALF-WAVELENGTH LINE CHARACTERISTICS

Let us analyze the voltage, current, and impedance relationships existing on a line that is short-circuited exactly one-half wavelength distant from the source. As shown in Fig. 2-13, maximum current flows at the short circuit and the voltage and impedance are at minimum value. A quarter wavelength back from the short circuit, in the direction of the source, the voltage and impedance are high, but the current is at a minmum. Note that only one-quarter wavelength of the line is shown in Fig. 2-13. At the source end of the line, exactly one-half wavelength from the short circuit, the line current, voltage, and impedance conditions are duplicates of those existing at the short circuit, and the

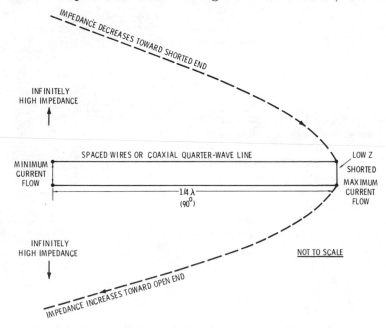

Fig. 2-13. Impedance characteristic of shorted quarter-wavelength line.

source also "sees" a short circuit at the line input. If the half-wavelength line is open-circuited exactly one-half wavelength distant from the source, the generator (transmitter) will "see" an open circuit at the line input. For example, if we terminate a 52-ohm RG-8/U line with 100 ohms at the half-wave point, the transmitter will "see" a 100-ohm load at the line input. If the half-wavelength is terminated with a capacitor, the transmitter will "see" a capacitive reactance equal to the X_C of the capacitive load; if the half-wave line is terminated with an inductance, the transmitter will "see" an inductive reactance equal to the X_L of the inductive load. It is obvious that the half-wavelength line will repeat the input whatever conditions are found at the other end. This characteristic is especially useful in antenna impedance measurements, and in the design of power-divider and phasing systems. The half-wavelength line will be discussed in more detail in the next chapter.

2-9. QUARTER-WAVELENGTH LINE CHARACTERISTICS

The quarter-wavelength line has perhaps more applications than the half-wavelength line. As shown in Fig. 2-13, when a quarter-wavelength line is short-circuited at the far end, the transmitter "sees" *not* a short circuit, but an open circuit. The half-wavelength line repeats its termination impedance at the input; the quarter-wavelength line *inverts* the load impedance. Thus, in a lossless quarter-wavelength transmission line shorted at the far end, the line input will present an infinitely high impedance to the transmitter output circuit. Since the half-wavelength and quarter-wavelength lines are actually a half-wavelength and a quarter-wavelength long at some particular frequency, at twice that frequency the half-wavelength line will be *two* half-wavelengths (a full-wavelength) long and the quarter-wavelength line will be *one* half-wavelength long. At *two times the fundamental frequency and at all even multiples thereof*, a quarter-wavelength line short-circuited at the far end will present a short-circuited input to the transmitter. At the fundamental frequency, however, the short-circuited quarter-wavelength line will present a *very* high impedance line input at the transmitter.

Fig. 2-14 illustrates a short-circuited quarter-wavelength line section connected across the main transmission line between the transmitter and the antenna. This quarter-wavelength *trap* section may be connected across the main line at any point from the transmitter to the antenna feedpoint and, when operated at the frequency at which the trap is one quarter-wavelength long, it will have no effect whatsoever on the function of the antenna system. At the second, fourth, and other *even* harmonics, however, the quarter-wavelength *stub* will reflect a short circuit across the main transmission line and will effectively suppress even-harmonic radiation from the antenna system. Stubs used to suppress harmonic radiation are usually referred to as linear traps.

Since the quarter-wavelength line characteristics are such that the load impedance between the two ends is *inverted,* it may be used as an *impedance transformer*. As an example, let us suppose that we wish to feed rf power to a quarter-wavelength, single-element vertical antenna system operated against a ground radial system. The feedpoint impedance at the base of the antenna, using an adequate system of radials, will be approximately 36 ohms. The only coaxial cable available in this hypothetical case is RG-

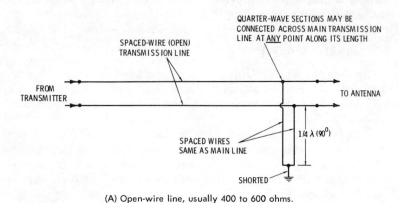

(A) Open-wire line, usually 400 to 600 ohms.

(B) Coaxial line, usually 50 or 75 ohms.

Fig. 2-14. Shorted quarter-wavelength line used as a trap.

11/U, which has a surge impedance of 75 ohms. If we connect the RG-11/U cable from the transmitter directly to the base of the antenna, there will be a mismatch of 75/36 or a ratio of 2.08:1. The swr on the RG-11/U line will be approximately 2:1. On the lower-frequency hf bands, a 2:1 swr is not considered to be extremely serious, but at the higher-frequency vhf and uhf bands, considerable power loss might result.

In order to match the 75-ohm transmission line to the 36-ohm antenna feedpoint, we are going to need some kind of impedance transformer which will step down the 75-ohm line impedance so that it "looks" like 36 ohms to the antenna. There are several matching devices that can be used, such as the pi and L-networks, toroidal-core impedance-stepdown transformers (to be discussed later), and the linear quarter-wavelength section. If we connect a quarter-wavelength piece of transmission line *with the correct surge impedance* between the far end of the 75-ohm line and the 36-ohm antenna feedpoint, the transmitter and main transmission line will "see" a 75-ohm resistance termination at the end of the RG-11/U cable, and the antenna will "look" into a 36-ohm termination at its feedpoint. The surge impedance (Z_0) value for the quarter-wavelength matching section is calculated from:

$$Z_0 = \sqrt{Z_1 Z_2}$$
$$= \sqrt{75 \times 36}$$
$$= \sqrt{2700}$$
$$= 51.96 \text{ or } 52 \text{ ohms}$$

where,

Z_0 is the required characteristic impedance of the quarter-wavelength matching section,

Z_1 is the impedance of the RG-11/U line,

Z_2 is the impedance of the antenna feedpoint.

In this case, RG-8/U coaxial cable has a characteristic impedance of 52 ohms. So, if we insert a quarter-wavelength section of RG-8/U cable between the antenna feedpoint and the far end of the RG-11/U line as shown in Fig. 2-15A, the system will, for all practical purposes, be almost perfectly matched. The swr in the RG-11/U line, which may be of any convenient length, will be about 1:1 ratio. The swr in the quarter-wavelength RG-8/U section will be about 2:1, but the power loss in the matching section will be *extremely low* because of its short length.

Fig. 2-15B shows a quarter-wavelength section of spaced line being used to match a 300-ohm ribbon line to a 600-ohm spaced-wire line. The required characteristic impedance (Z_0) for the quarter-wavelength spaced-wire line is calculated from:

$$Z_0 = \sqrt{Z_1 Z_2}$$
$$= \sqrt{300 \times 600}$$
$$= \sqrt{180,000}$$
$$= 424 \text{ ohms}$$

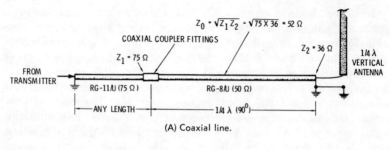

(A) Coaxial line.

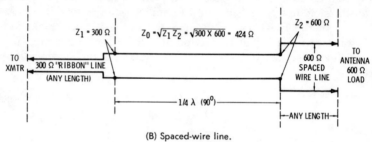

(B) Spaced-wire line.

Fig. 2-15. Quarter-wavelength sections used as matching transformers.

where,

Z_0 is the required characteristic impedance of the quarter-wavelength section of spaced-wire line,

Z_1 is the impedance of the 300-ohm ribbon line,

Z_2 is the impedance of the 600-ohm spaced-wire line.

Therefore, we need a quarter-wavelength section of spaced line with a characteristic surge impedance (Z_S) of 424 ohms. The characteristic surge impedance of a spaced-wire line can be determined by the equation:

$$Z_S = 276 \log \frac{2D}{d}$$

where,

Z_S is the characteristic surge impedance of the spaced-wire line,

D is the spacing between the conductors, center to center,

d is the diameter of the conductors, expressed in the same unit as D.

In the preceding example, we used a quarter-wavelength line to match a lower-impedance antenna feedpoint to a higher-impedance transmission line. The reverse of this procedure also may be used. Radio amateurs often use *long-wire* antenna systems operated at multiples of half or quarter wavelengths. Antenna systems operated at an *odd number* (3, 5, 7, 9, etc.) of half wavelengths have a current loop (maximum) at the center and also at one-quarter wavelength from each end, as shown in Fig. 2-16. The radiation resistance (impedance) at these current loops varies from about 80 to 130 ohms, depending upon the number of half waves on the antenna, but may be roughly estimated to be in the vicinity of 100 ohms for most installations. Suppose that we have a quantity of RG-8/U (52-ohm) line on hand. How can we match the 50-ohm line to the 100-ohm antenna feedpoint? Again, there are many devices available, but the simplest and easiest method probably will be to use the quarter-wavelength impedance-match-

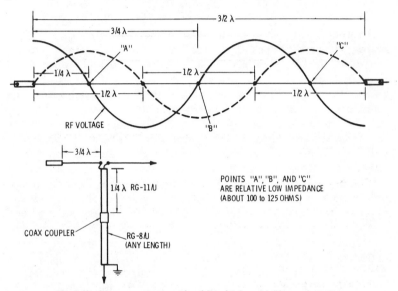

Fig. 2-16. Antenna systems with odd multiples of half wavelengths.

ing transformer. In this case, the Z_0 of the matching section will be:

$$Z_0 = \sqrt{50 \times 100}$$
$$= \sqrt{5000}$$
$$= 70.7 \text{ or } 71 \text{ ohms}$$

The surge impedance of RG-11/U is nominally 75 ohms and a quarter-wavelength section of this cable connected in series with the RG-8/U main line and the antenna feedpoint will give a very low swr on the 50-ohm line back to the transmitter. When matching horizontal antenna systems to coaxial transmission lines, the match can sometimes be greatly improved by changing the height of the antenna a few feet up or down, while observing an swr meter or other indicator in the main line to the transmitter.

The physical length of a half-wavelength section of RG-8/U or RG-11/U with standard dielectric A (not foam) can be determined as follows:

$$\text{one-half wavelength (ft)} = \frac{324.72}{F\,(\text{MHz})}$$

To determine the physical length of a quarter-wavelength section of RG-8/U or RG-11/U, use the following formula:

$$\text{one-quarter wavelength (ft)} = \frac{162.36}{F\,(\text{MHz})}$$

Another method of eliminating standing waves from a line with an impedance that does not match its load is by the use of *matching stubs*. Matching stubs consist of short sections of transmission line cut so that they will present a desired impedance to the line and to the load. When a matching stub is connected across the main transmission line, *at the proper point,* it cancels the line reactance due to the existing mismatch. The net result is that the line from the transmitter to the point where the stub is connected will be nonresonant with a very low swr. Standing waves will still exist on the stub and on the line between the point where the stub is connected and the antenna. Because the impedance value along a quarter-wavelength line, shorted at one end and open at the other, varies from practically zero at the shorted end to infinity at the open end, any line and any impedance can be matched at some point on the stub. Mathematical analysis of stub action and calculations of stub lengths and locations is a very complex process and would be of interest to only a few readers of this book. In most cases, the connecting point and length of both shorted and open matching stubs may be determined from the graphs in Fig. 2-17. After the stubs are installed, the line swr or the impedance at the point on the stub where the line is attached should be measured and the stub length adjusted for the lowest line swr or the correct impedance at the stub feedpoint. This subject will be treated in more detail later.

2-10. LINEAR POWER DIVIDERS AND PHASING NETWORKS

In phased antenna arrays, where all of the elements are driven, it is very important to control the relative amplitudes, as well as the phase, of the rf currents supplied to the various elements. In most amateur-radio antenna systems, equal values of rf currents are supplied to the different elements; in commercial installations, particularly am broadcast-station antennas, the amplitude of the current in each leg of the array may be different from that of the others. Since the patterns are generally fixed and closely controlled, both the current and phase may be adjusted to produce the desired directivity and pattern shape. The

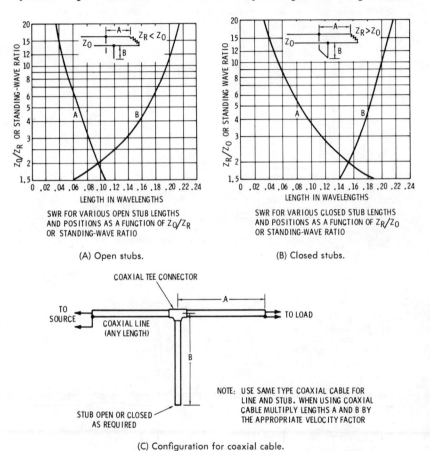

(A) Open stubs.

SWR FOR VARIOUS OPEN STUB LENGTHS
AND POSITIONS AS A FUNCTION OF Z_0/Z_R
OR STANDING-WAVE RATIO

(B) Closed stubs.

SWR FOR VARIOUS CLOSED STUB LENGTHS
AND POSITIONS AS A FUNCTION OF Z_R/Z_0
OR STANDING-WAVE RATIO

NOTE: USE SAME TYPE COAXIAL CABLE FOR
LINE AND STUB. WHEN USING COAXIAL
CABLE MULTIPLY LENGTHS A AND B BY
THE APPROPRIATE VELOCITY FACTOR

(C) Configuration for coaxial cable.

Courtesy Collins Radio Co.

Fig. 2-17. Graphs for determining lengths and positions of open and closed matching stubs.

portion of the transmission-line system that controls the amplitude of the currents fed to the individual elements is called a *power divider*. The portion of the line system that controls the relative phase (timing) of the currents supplied to the different elements is called a *phasing network* or *harness*. In addition to the power distribution and relative phasing, the various impedance values existing throughout the system must be matched to their respective loads. All of the three factors will interact to some extent and will require adjustment after the array is completed. An additional function of the power divider is to prevent interaction between the different elements of the array *through the feeder system.* For proper gain and directivity, the isolation between elements should be at least 25 dB.

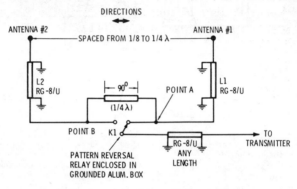

(A) Produces reversible cardioid pattern only.

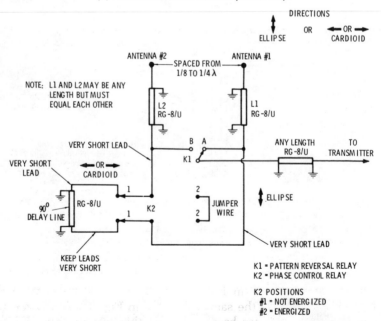

(B) Produces reversible cardioid pattern or elliptical pattern.

Fig. 2-18. Simplified vertical two-element antenna array.

Most phasing networks are designed so that one or more elements of an array are excited at a time later than that of some other element used for reference. The array shown in Fig. 2-18 is a simplified vertically polarized, phased antenna system. In the arrangement shown in Fig. 2-18A, the array is designed to transmit a cardioid pattern in either of two directions, along a line drawn through the plane of the two antennas as indicated by the heavy arrows. The pattern radiated from the array illustrated in Fig. 2-18A is a cardioid in shape and its direction is controlled by the pattern-reversal relay, K1. Cardioid patterns are shown in Figs. 2-19A and B.

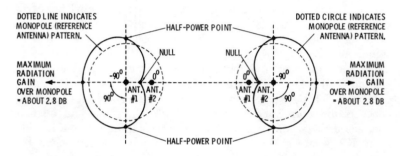

(A) Pattern with relay K1 in position 2 and relay K2 in position 1.

(B) Pattern with relay K1 in position 1 and relay K2 in position 1.

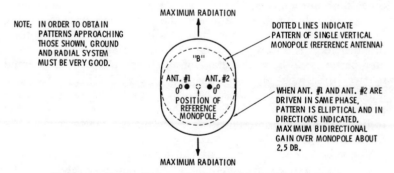

(C) Pattern with relay K1 in position 1 or 2 and relay K2 in position 2.

Fig. 2-19. Directional characteristics of simple two-element array.

In the array shown in Fig. 2-18B, the antenna elements and their arrangement is the same as shown in Fig. 2-18A. However, a second relay, K2, has been added. In this system, relay K1 is the pattern-reversal relay and K2 is the phase-control relay. Relay K1 is a spdt coaxial type used to connect the transmission line from the transmitter to either point A or point B. When relay K1 is not energized, the main transmission line is connected to point A and through line L1 to antenna No. 1. In Fig.. 2-18A, the signal passes through the 90° delay line to point B and through line L2 to antenna No. 2. Since lines L1 and L2 are *exactly* equal in length, the signal at antenna No. 2 will be delayed 90° with reference to the signal at antenna No. 1 by the delay line. The signals passing through lines L1 and L2 will also be delayed but the delay in the two lines is equal and does not affect the relative phase of the signals at the two antennas. When relay K1 is energized, the transmission line from the transmitter will be connected to point B and antenna No. 2 becomes the reference (0°) antenna. To reach antenna No. 1, the signal must pass from point B through the 90° delay line to point A. Now, the signal arriving at antenna No. 1 is delayed by 90° with respect to the signal at antenna No. 2. Maximum signal radiation will occur along the line drawn through the two antennas and in the direction of the lagging-phase element.

In Fig. 2-18B, when relay K2 is not energized, the armature contacts are connected across the 90° delay line and also are connected to points A and B at the contacts of relay K1. When relay K2 is energized, its armature contacts are connected to a jumper and the 90° delay line is removed from the circuit. In this position, points A and B are connected together and the signal from the transmitter is applied simultaneously to both antennas. Since both antennas receive the signal at exactly the same time, there is no relative phase difference between the currents flowing in the two antennas and maximum radiation will occur along a line drawn at right angles to the plane of the antennas. The radiated pattern will be in the shape of an ellipse and will be radiated in two directions simultaneously (Fig. 2-19C). When the two antennas are driven *in phase,* relay K1 may be left in either position since it now has no effect on the pattern. When relay K2 is in the cardioid position, however, relay K1 is used to reverse the direction of the cardioid pattern.

The relays should be mounted in a waterproof aluminum box and all external connections should be made with standard coaxial plugs and receptacles. The relay box should be grounded to an *earth* ground. The connecting leads from the coaxial fittings to the relays and the wiring between the relays should be heavy, short, and direct. If long leads with kinks or bends are used, an additional phase shift will be caused by the lead inductance. This stray phase shift may be sufficient to deteriorate the performance of the array. Such phase shift will particularly affect the front-to-back pattern ratio.

2-11. SUMMARY OF TRANSMISSION LINES, POWER DIVIDERS, AND PHASING NETWORKS

There is no intention of making this book a treatise on transmission lines. However, there are certain basic principles involved in making impedance, phase, swr, and other measurements and adjustments in multielement antenna systems that must be clearly understood. Unless the reader has had considerable experience in the design and adjustment of complex antenna systems, he would be well advised to experiment with one of the simpler systems, such as that just described before attempting to design, construct, and adjust a three- or four-element phased array. There is a wealth of published data in radio-amateur literature. Also, a great amount of essential basic information on the design and adjustment of antenna systems was published in the technical texts and literature of the 1940s and 1950s. These publications are generally available at most large public libraries.

Most radio amateurs design and construct antennas by calculating cable and element lengths from published formulas and tables. While antennas designed by this method almost always function to some extent, their performance can usually be improved by proper adjustments based on accurate measurements of the system characteristics. The principal reason for the lack of measurement data, other than that of impedance, is the almost total absence of accurate test instruments at the average amateur-radio station. The adjustment of a complex directional array with nothing more than a resistance-type swr bridge and a monimatch unit would be a nerve-wracking, time-consuming process to say the least. In addition, the adjustment of the phasing harness of a multielement array by clipping coaxial line lengths can be an expensive procedure at current prices. Relatively inexpensive and simple test equipment, constructed by the radio amateur himself, will enable him to be *sure* that his antenna system is performing at peak efficiency. The test equipment described in the following chapter was designed and constructed over a period of several years. Most of it was built from "junk-box" parts found in many ham-shacks.

SELF-EXAMINATION

Here is a chance to see how much you have learned about transmission lines. These exercises are for self-testing only. Answer true or false.

1. A 12-volt battery is suddenly connected across a length of RG-8/U, 53-ohm, coaxial cable. The cable is instantly charged to a potential of 12 volts dc.

2. A radio wave travels through space at the velocity of light. Its velocity traveling down a coaxial line is the same as in space.

3. During the time that a transmission line is being charged, an actual electron current flows through the dielectric (insulation).

4. A section of RG-8/U coaxial cable is 10 meters long (measured). Its electrical length is also 10 meters long.

5. A section of RG-8/U coaxial cable is 22 feet, 6 inches long. The surge impedance is 52 ohms. Therefore, a 45-foot section of RG-8/U will have a surge impedance of 104 ohms.

6. We need a coaxial cable with a surge impedance of 26 ohms. We can make up such a cable by connecting two equal-length sections of RG-8/U in parallel.

7. A transmission line connected between two antennas is exactly one electrical wavelength long. If the main transmission line from the transmitter to the first antenna is one-half wavelength (180°) long, the two antennas are driven in phase.

8. The swr expresses the degree of mismatch between the transmitter output terminal and the transmission line.

9. An antenna used *only for receiving* should be matched to the transmission line at the antenna feedpoint.

10. An antenna used *only for transmitting* should be matched to the transmission line at the antenna feedpoint.

CHAPTER 3

Antenna Measurements and Test Equipment

The contents of this chapter will introduce the reader to the various types of test equipment necessary for an accurate determination of the important factors in an operating antenna system, whether it be a simple vertical rod or a complex phased array. In order to obtain accurate data on the different characteristics of a system, several different pieces of test equipment will be required. For example, to check the radiation pattern and to measure the field strength of an antenna, we must use a *field-intensity meter*. To measure the feedpoint impedance we use an *impedance bridge,* either a resistance bridge or an R-X bridge. The transmission-line swr and the line-to-antenna matching is checked and adjusted by means of a *standing-wave indicator*. In addition, it is sometimes necessary to check transmitter operation, or make comparison tests on the system, without allowing the antenna to radiate rf energy. In this case, the actual antenna is replaced by a pure resistance load known as a *phantom* or *dummy antenna*. The dummy antenna is useful for making *rf power* measurements, either at the transmitter output or at the end of the transmission-line system.

3-1. LOW-LEVEL SIGNAL GENERATOR

All of the instruments to be described in this chapter require some kind of radio-frequency signal source in order to present meter readings, or other visual indications. When these indications are correctly interpreted, they will enable the radio amateur to determine the actual operating conditions within the system. Many radio amateurs use their transmitters as signal generators when making antenna adjustments. This is not only a very dangerous practice, but the excessively strong signal throughout the system makes the use of sensitive test instruments impractical. The rf output from a transmitter is measured in watts; the rf output from a signal generator is usually measured in *millivolts*. The rf output signal should be capable of being attenuated practically to zero level if desired, and the frequency should be very stable, preferably crystal controlled. Obviously, a test signal of such a low amplitude will require extremely sensitive indicating meters. All of the instruments described use either low-range microammeters or a vtvm with low-voltage scales to visually indicate the circuit conditions. The author uses a vtvm which has low-voltage dc scales, beginning with 0.5 volt. Any of the low-cost, imported voms with low-current and low-voltage scales are excellent for this purpose. The small testers with 20,000-ohms-per-volt sensitivity have a 0–50-microampere current range which is very sensitive as a visual indicator.

Fig. 3-1 shows a small low-level signal generator for the 40-meter band constructed mostly from "junk-box" components. The schematic diagram for the generator is shown in Fig. 3-2. Although this generator uses a 6CL6 vacuum tube in a standard crystal-oscillator circuit, a transistorized oscillator can be used if desired. The rf output from the generator should be in the order of one volt rms measured under load. The output circuit should be provided with an attenuator so that the rf signal can be adjusted to any desired level. The test signal is usually adjusted for full-scale deflection of the indicating meter before connecting the instrument to the circuit under test or adjustment. The generator shown here is suitable for all our purposes except for the phase measurements. An rf signal of about 75 volts peak-to-peak is required to deflect the beam of the small vectorscope used for the visual phasing adjustments.

The signal generator shown in Fig. 3-1 is crystal-controlled. The dial on the front panel is the attenuator. The use of a grid-dip oscillator as a signal source is to be discouraged, particularly for the phasing adjustments. In the first place, the grid-dip signal is relatively unstable and, unless monitored constantly by a receiver, the exact frequency of operation cannot be determined with any degree of accuracy. In the second place, there is no provision for attenuating the grid-dip oscillator signal. Some of the tests described here require attenuation of the signal. The instrument shown in Fig. 3-1 is used mostly for checking the author's

Fig. 3-1. Low-level, fixed-frequency signal generator.

40-meter array. In this particular case, all adjustments are made at 7.2 MHz. In this instrument, the plate circuit is slug-tuned and the output is taken by means of a single turn of hookup wire as shown in the schematic. A 100K, 2.5-watt potentiometer in series with the link coil acts as an attenuator. The instrument is completely shielded against signal radiation from the circuit components. When the generator is placed on top of the station receiver and its output cable and connectors are removed, the signal is barely audible when the receiver is tuned to the generator frequency. The rf output from the front-panel coaxial terminal is fed through a shielded 50-ohm coaxial cable to the circuit under test. The power-supply circuit is also shown in Fig. 3-2. This signal generator may be used on the 20-meter (14.2 MHz) and 15-

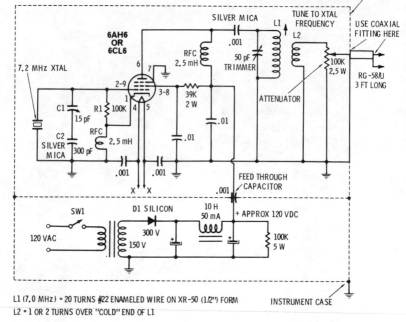

Fig. 3-2. Schematic diagram of low-level signal generator shown in Fig. 3-1.

meter (21.3 MHz) bands by changing the crystal and tank circuit accordingly.

3-2. RESISTANCE IMPEDANCE BRIDGE

Two basic resistance-bridge circuits are shown in Fig. 3-3. The circuit shown in Fig. 3-3A is excellent, particularly if measurements are to be made well into the vhf region. However, this circuit requires a different variable capacitor for C1 and C2. These capacitors are not ordinarily found in the average amateur's junk box and, while still available, are fairly expensive. The simple resistance bridge shown in Fig. 3-3B is satisfactory for use at all frequencies below 30 MHz if carefully laid out and constructed to reduce coupling between the resistors that form the bridge arms.

The practical bridge circuit shown in Fig. 3-4 is constructed in an aluminum box 4¼ inches long, 2¼ inches wide, and 3 inches high. The input and output coaxial fittings are mounted on the sides. The component layout and schematic diagram are shown in Fig. 4-5. Series resistor R3 is made up of five selected 270-ohm, 1-watt carbon resistors which are placed in parallel to give a nom-

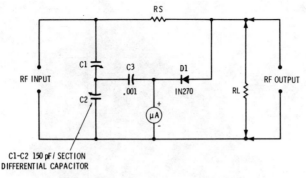

(A) Circuit for frequencies in the vhf range.

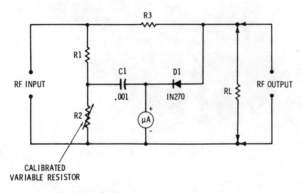

(B) Circuit for frequencies up to 30 MHz.

Fig. 3-3. Two basic resistance-bridge circuits.

inal 52-ohm resistance rated at 5 watts. By measuring a quantity of 270-ohm, 1-watt resistors with an accurate ohmmeter, five resistors can be selected within the standard tolerance which, when connected in parallel, will measure exactly 52 ohms. The leads of the paralleled resistors must be kept as short as possible and soldered together. When properly connected, the paralleled resistors will just fit between the terminals of the coaxial fittings with the leads less than one-quarter inch long. Resistance R2 consists of two 2-watt, 100-ohm carbon resistors, selected to measure 53 ohms when connected in parallel. Resistance R2 should be mounted at right angles to R3 to prevent rf coupling. Potentiometer R1 is an Allen Bradley type-J carbon linear control rated at 2.5 watts. The meter resistor, R4, is a 1200-ohm, 2-watt carbon type. The value of R4 is not critical. However, all of the resistors should have good rf characteristics. The Ohmite Little Devil series or Allen Bradley types are recommended. Avoid cheap "bargain-package" resistors. The potentiometer is fitted with a 270° dial scale. The two binding

Fig. 3-4. Practical rf resistance bridge.

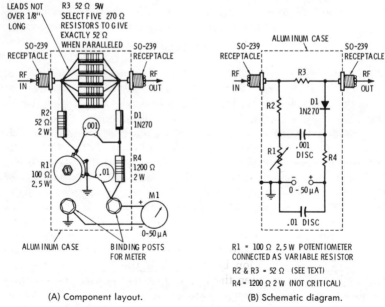

(A) Component layout.
(B) Schematic diagram.

Fig. 3-5. Component layout and schematic diagram for practical rf resistance bridge.

posts at the top are for connecting a 0–1-mA dc milliammeter. (For use with 0–.5-volt dc scale on vtvm, connect a 220,000-ohm resistor across binding posts.)

To test the bridge, connect the rf signal generator to the bridge input terminal. With an rf probe connected to the vtvm, measure the voltage at the *unterminated* bridge output terminal and adjust the level of the rf signal until the meter reads full scale on the 0–.5-volt range. If the 0–1-mA dc meter is used, follow the same procedure and adjust the rf level for full-scale deflection. Then, terminate the output terminal of the bridge with a 52-ohm carbon resistor. Keep the leads as short as possible. Rotate the bridge dial for minimum indication on the meter. At one point on the dial scale, the meter should "dip" to zero indication. If it does not, either the terminating resistor is not a pure resistance (has lead inductance) or there is coupling between the resistances and the diode which form the bridge. Change the position of the bridge resistors, or the diode, slightly or reduce the lead length of the terminating resistor until the meter reads zero at the bottom of the null. Note the dial reading at which the null is obtained. This is the 52-ohm calibration point. Now, terminate the bridge with other resistor values (75 ohms, 36 ohms, etc.) and note the dial reading for each null. After you have calibrated the bridge in the manner described above, it is ready for use. To become familiar with the operation of the bridge, we will measure the base *resistance* of a simple quarter-wave vertical antenna.

The signal generator and indicating meter are connected to the bridge, as before. With the bridge output terminal *open,* adjust the signal level for full-scale deflection on the meter. Connect the antenna under test to the bridge output terminal through a *short* length (6 to 8 inches) of coaxial cable. Rotate the bridge dial until a null is obtained. In most cases, the meter will not "dip" to zero at the null when the antenna is connected to the bridge. This condition indicates that the antenna feedpoint contains *reactance* as well as resistance. With this simple bridge circuit we are unable to determine whether the reactance present is *capacitive or inductive.* The usual procedure is to add either inductive reactance (inductor) or capacitive reactance (capacitor) to the antenna circuit and adjust the bridge for a null. If you have added a coil to the antenna circuit and the bottom of the null is closer to zero, the antenna is too short for the frequency in use. Continue to add inductance to the circuit, one turn at a time, until the lowest null indication is obtained. If the bottom of the null was higher in value when the inductance was added, the antenna is too long for the frequency. Add capacitance (preferably variable) to the antenna circuit, adjusting the capacitance value while rotating the bridge dial for the null. If the meter indication is at or near zero when nulled at some capacitance value, the antenna is *resonant* at the frequency of the signal generator and the *feedpoint impedance* (resistance) may be read from the calibrated bridge dial.

The value (in ohms) read from the bridge is the impedance of the antenna feedpoint. For the vertical quarter-wave antenna, this reading may be anywhere from 36 ohms to 60 ohms, or higher, depending on the ground losses. Assuming a perfect ground system, the feedpoint impedance at the base of a quarter-wave vertical should be about 36 ohms. Note that you have *tuned out the reactance* of the antenna and measured its feedpoint resistance. If *all* of the reactance has been tuned out, assuming a perfect ground system, the feedpoint impedance will be about 36 ohms. In order to match the feedpoint impedance to a standard 50- or 75-ohm coaxial line, we must design a matching network to be inserted between the coaxial line and the antenna feedpoint. The design of matching networks will be covered later.

3-3. RESISTANCE-REACTANCE (R-X) BRIDGE

The resistance bridge is very useful in determining the impedance of antenna feedpoints. However, antenna systems rarely present a pure resistance load to a coaxial line. This is particularly true in vertical arrays using large diameter aluminum pipe or tubing for the elements. When cut according to an antenna design formula, large tubing antenna elements tend to resonate at frequencies generally lower than that used in the formula, even when the length factor K is considered. The deviation appears to be more pronounced when the array is erected near buildings or in the vicinity of power lines, water tanks, etc. In spite of very careful calculations and measurements, practically every vertical antenna erected by the author during the past 20 years has resonated at a frequency lower than that used in the calculations.

An antenna element that is resonant at a frequency lower than the design frequency is too long and will contain an inductive reactance component (X_L) at the design frequency. If the antenna resonates at a frequency higher than the design frequency, it is too short and will contain a capacitive reactance component (X_C) at the design frequency. As discussed previously, when a resistance bridge is used to measure the base feedpoint impedance (resistance), an incomplete null indicates reactance but does not specify whether it is X_L or X_C. The antenna element must be taken down and adjusted to the proper length for resonance, or we must insert a reactance in the antenna circuit equal to and

(A) External view.

(B) Internal construction.

Fig. 3-6. The R-X bridge.

opposite in sign to that of the antenna reactance in order to produce a pure resistance at the antenna feedpoint. The first corrective procedure entails considerable work in taking down and reinstalling vertical elements, particularly for the 40-, 75-, and 160-meter bands. The second procedure may require much tedious experimenting before the reactance with the proper sign and value is found. The R-X bridge is designed to eliminate these shortcomings of the resistance impedance bridge. The R-X bridge shown in Fig. 3-6 will indicate the sign of the inherent reactance of the antenna element and will also give an approximate indication of the value of the added inductive or capacitive reactance required to produce a pure resistance feedpoint impedance. Now let us go back to the quarter-wavelength vertical antenna and measure its base impedance with the R-X bridge.

There is nothing radically different or new about the R-X bridge. The basic principles have been used for many years by broadcast engineers to adjust and match broadcast-station antennas. Instruments similar to that described here have appeared in amateur-radio literature over the past few years. Most amateurs are familiar with the purpose of the R-X bridge even though they may not thoroughly understand the principles involved in its use. The schematic diagram for the R-X bridge circuit is shown in Fig. 3-7. The instrument, which is designed for the 40-meter band only, uses the same circuit as that of the resistance bridge for measuring the antenna resistance, but contains an LC circuit (C1 and L1) which is *series-resonated* at the design frequency. In a series-resonant ac circuit, the X_L and X_C components cancel at resonance, resulting in a net reactance of zero ohms across the

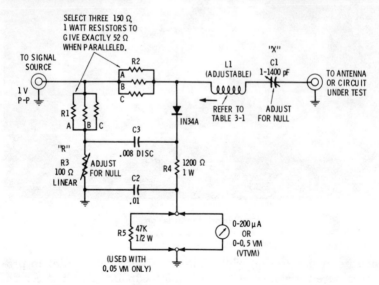

Fig. 3-7. R-X bridge for resistance and reactance measurements.

terminals of the resonant circuit. In the R-X bridge, the series-resonant circuit is connected in series with the output terminal of the bridge. When C1 and L1 are series-resonated at the design frequency and the R-X bridge is terminated by a pure resistance, the indicating meter should read at or very close to zero. The inductance, L1, is adjusted so that the capacitor, C1, resonates the tuned circuit to the design frequency when the capacitor plates are about half meshed (Fig. 3-6B). The variable capacitor is fitted with a zero-center dial calibrated from 0 to 100 in each direction from the center.

Referring to Fig. 3-6A, the R dial for adjusting the 100-ohm potentiometer (R3) is at the left. The X dial for adjusting variable capacitor C1 is located just to the right of the R dial. The screw adjustment at the bottom center adjusts the slug of coil L1. The upper SO-239 coax connector is for the rf input signal. The lower coax connector is connected to the antenna or transmission line that is being measured. The indicating meter is connected to the two binding posts located between the coax connectors. The winding data for coil L1 is given in Table 3-1.

Table 3-1. Winding Data for Coil L1 in Fig. 3-7

Band (Meters)	Diameter Coil Form (Inches)	Turns (Closewound)	Wire Size (Enameled)
75-80	½	23	30
40	½	15	24
20	½	6	20
15	½	4	20
10	½	3	18

All coils are wound on National XR-50 powdered-iron, slug-tuned forms. Adjust number of turns and spacing until resonance at the test frequency occurs within range of slug adjustment when the dial is set at center zero. Coils may be mounted in plug-in forms for convenience in changing from one band to another. Do not use a bandswitch to change coils.

To calibrate the reactance dial, connect a 52-ohm carbon resistor across the output cable. Now, rotate resistance dial R for the null. It is likely that the depth of the null will not reach zero on the meter. Turn reactance dial X until the deepest null is obtained. By careful alternate adjustments of the R and X dials, the meter should read zero at the bottom of the null. If it does not, there may be intercoupling between the bridge resistors, or there may be reactance in the resistor leads and clip leads at the end of the output cable. An absolute-zero null indication is of no great importance unless it is desired to determine the absolute values of the resistive and reactive components of the antenna under test. Terminate the output cable with different values of resistance in the same manner as for the simple resistance bridge, rotate the R dial for the null, and record the dial settings. Always tune the X dial for greatest null depth.

In order to calibrate the X dial, it is necessary to add known values of X_L and X_C in series with the terminating resistance across the output cable. If the X dial was set at zero center and the LC circuit was resonated by adjusting the slug of L1 for deepest null with the output cable terminated in a resistance, the addition of either a coil or a capacitor in series with the termination resistance will necessitate a readjustment of the X dial to obtain resonance again. The resonant point on the X dial now will be either above or below zero, depending upon the sign (X_L or X_C) of the added reactance.

Most amateur-radio technicians are not interested in analyzing the complex R and X components at the antenna feedpoint, but merely wish to determine whether to add inductance or capacitance to the antenna circuit in order to bring it to resonance at the design frequency. If the terminating impedance contains *inductive reactance,* then *less capacitance* (higher X_C) is required from C1 to establish the null. If the terminating impedance contains *capacitive reactance,* then *more capacitance* (lower X_C) is required from C1 to establish the null. Thus, if less capacitance from C1 is required to establish the null when measuring the antenna feedpoint impedance, the antenna contains inductive reactance (too long) at the design frequency. If more capacitance from C1 is required to establish the null, the antenna contains capacitive reactance (too short) at the design frequency. An antenna with X_L at the operating frequency may be effectively "shortened" electrically by adding X_C to the antenna circuit. An antenna with X_C at the operating frequency may be effectively "lengthened" electrically by adding X_L to the antenna circuit. When the X_L of the antenna is equal to the antenna X_C, the two reactive components cancel and the antenna is said to be *resonant* at the design or operating frequency. Note that we have not tuned or matched the antenna to any specific impedance value. Impedance matching will be discussed later.

3-4. PHASE MEASUREMENTS WITH THE QUADRASCOPE

In most cases, the radio amateur will obtain design information from the radio-amateur literature, handbooks, magazines, or other publications. In simple balanced arrays (not operated against ground), particularly horizontally polarized antennas, phasing is a lesser problem as compared with complex vertical arrays which are fed and phased with coaxial lines almost exclusively. If a vertical array is constructed, fed, and phased strictly from published data, the chances are that it will work reasonably well and may exhibit considerable improvement over the performance of a single quarter-wave vertical monopole. However, the improved performance may be due to the lowered vertical radiation angle of the array rather than its supposed directional characteristics. The performance of any antenna system should not be evaluated by the results of one evening's operation which might be misleading due to unusually good or poor band conditions. In most cases, operation over a period of at least a month is desirable in order to obtain any useful data.

Generally, the radio amateur will follow a pattern when checking the performance of a new array. The first step is to determine whether or not the system will load the transmitter. If it does, and a contact is established, the second step will be to actuate the various switches and relays while asking for comparative signal strength reports for the various switch positions. If the array is being operated on the 40-meter band and the contact is less than 1000 miles away, the amateur may be astounded by the fact that when the switches are thrown to a position where the signal *should increase,* the signal strength actually *drops.* This condition is now more prevalent on the 40-meter band when the distance between the two amateur stations is 400 miles or less. In most cases, the signal from the station using the array will be very strong (S9 plus). What is wrong? It is entirely possible that the array may be performing normally. It must be remembered that a vertically polarized, phased, driven array is primarily a *long-distance* antenna system. The major lobe in the vertical plane is radiated at a very low angle—usually around 28° to 30° with respect to the earth's surface, depending somewhat upon the effectiveness of the ground-radial system. This does not mean, however, that all of the radiation from the array takes place at the low vertical angle of the major lobe. The radiation pattern of any antenna system, vertical or horizontal, contains minor lobes at various vertical angles. It is also possible that some metal object, such as a guy wire or a power-line wire, may be excited by the strong radiation field of the array and is reradiating a high-angle wave. Today, it is not uncommon for an amateur station to feed power levels of several hundred watts to the antenna system and, as a result, even the minor-lobe signals may be very strong.

After much experimentation and testing with other amateurs at varying distances, the author has concluded that the strong signals which are received at relatively short distances from the array, and which do not correspond to theoretical expectations of the array phasing, are due to *high-angle minor-lobe radiation.* In order to check the directivity characteristics of an array, the reporting station should be at a considerable distance (over 1000 miles when using the 40-meter band) from the transmitting station. If the performance does not follow the theoretical expectations, the array may not be properly phased or the array pattern may be distorted due to reradiation from nearby metallic objects.

If possible, it is advisable to plot the field pattern of the array with a field-strength meter before making radical changes in the array or its phasing system. If it is suspected that parasitic secondary radiation is taking place from nearby metallic objects, have someone ground these objects while observing the field-strength meter. (**Do not attempt to ground power-line wires.**) If the reading is affected, the object is probably being excited parasitically and is reradiating. Sometimes reradiation from a metallic object can be detected by bringing the field-strength meter in the vicinity of the object. The author has had considerable difficulty with a water-pressure tank and aluminum roofs on buildings near his 40-meter array. All of these objects were grounded which resulted in some improvement but did not entirely eliminate the pattern distortion. In this particular case, when transmitting toward the east, stronger signal reports are given by stations located in the southern half-section of the major lobe. The field-strength patterns should be plotted *before and after* any major changes in the array are made. A comparison of the before and after plots will often yield a great deal of valuable data.

Phasing errors are most likely to occur when different types of switching relays are used in the phasing system or there is coupling between the relays or other components of the system. The phasing lines themselves must be cut precisely for the desired relay—most likely 90°, 135°, or 180° degrees. Methods of measuring the *electrical length* of phasing lines, using the R-X bridge, will be discussed later. All connecting leads between the coaxial line terminals and the relay contacts should be short and direct. All corresponding leads in each switching circuit should be of the same length and the *lead dress* in each circuit should be as nearly alike as possible. It is recommended that the switching relays and associated wiring be separated from each other, or at least be placed in separate shielded compartments. It is not advisable to roll up coaxial lines and place them in the vicinity of each other or bury them in a common trench in the ground. In at least two instances, the front-to-back ratio of an array is known to be affected when the coaxial phasing lines were buried in the earth along with the radials. The exact cause of the problem is not known, but it is suspected that ground-return current loops were formed when the cables were buried in the vicinity of the radials.

Another potential source of phasing troubles is the impedance-matching networks between the coaxial lines and the antenna feedpoint. In order to obtain maximum radiation efficiency from the system, the antenna feedpoint impedance must be matched to the surge impedance of the transmission line. This is usually accomplished by means of a broad-band, toroidal, impedance-matching transformer or by a simple L-network. In all cases, the matching network will cause a phase shift to occur in its leg of the array. To minimize the effect on the overall system, the matching networks in each leg of the system should be identical so that their phase shifts will be equal. However, there may be a slight difference in the individual characteristics of the various elements, requiring a slight difference in the matching-network values.

Dozens of tests and measurements on the author's arrays indicate that the most common symptom of improper phasing is a poor front-to-back ratio and the lack of null depth in cardioid-type patterns. As an example, the author's station, located in northern California and operating on the 40-meter band, established contact simultaneously with a station in Rochester, New York and another station in Hawaii. Although the pattern at that time was a cardioid being transmitted east, the signal in Honolulu was only slightly weaker than the east-coast mainland signal. Although the phasing lines had been calculated and measured physically, *electrical measurements* with a bridge indicated that they were incorrect. This array, before correction, also had poor rejection of foreign broadcast stations received in the pattern null. If the array is properly phased, there should be a very noticeable signal drop when the pattern is switched to place a strong interfering signal in the pattern null.

At this point, the reader may wish to again read Section 2-3— *Phase Shift in Transmission Lines* before continuing with our present discussion. You will note that we stated: "If the cable characteristics are uniform along its length (no sharp bends or dents) and it has been terminated in a *pure-resistance load* equal to its surge impedance, the electrical one-wavelength line may be divided into electrical degrees." The term pure-resistance load does *not* necessarily mean that the antenna load is a *physical* resistor. It *does* mean, however, that coaxial feed line must be matched at the antenna feedpoint so that the antenna element presents a resistance (simulated) load or termination to the coaxial cable. It is a waste of time and energy to attempt to make phasing adjustments in any antenna system before the line-to-antenna impedance matching is completed. That is, line and antenna must be matched before phasing adjustments are made.

The formulas for calculating 90° and 180° sections (one-quarter and one-half wavelengths, respectively), as given in Section 2-9, are valid for most standard 50- and 75-ohm coaxial cables in new condition with type-A dielectric. If the phasing lines are to be cut from used or surplus cables, the phasing-line length should be determined by *electrical* measurement rather than by strictly physical measurement. The electrical length of a cable may be determined by the use of the R-X bridge or by means of the quadrascope (vectorscope) to be described later.

From Section 2-8, we learned that a half-wavelength line terminated in a given resistance value will always repeat its termination value at the input end. To check a half-wavelength line with the R-X bridge, calculate the *physical* length of the line from the formula given in Section 2-9. When cutting the line, however, make it at least a foot *longer* than the calculated length. Terminate one end of the line with a fixed carbon resistor having a resistance value about twice the surge impedance of the coaxial cable under test. For RG-8/U and similar 50-ohm cables, terminate the line with a 100-ohm resistor. For RG-11/U and similar 75-ohm cables, use a 150-ohm terminating resistor. *Do not use a terminating resistor equal to the surge impedance of the line.* Before connecting the resistor across the cable, use the resistor to temporarily terminate the output terminals of the bridge, and adjust the bridge R and X dials for the lowest null indication. Now, use the *same* carbon resistor to terminate the far end of the cable. Connect the other end of the cable to the R-X bridge output terminal and observe the reading on the bridge indicating meter. The meter will probably read up-scale. Do *not* readjust the bridge R and X dials. Starting at the far end of the cable, clip off an inch or so of the cable and reconnect the terminating resistor across the cable. Continue to shorten the cable and reconnect the terminating resistor until the bridge meter reads exactly the same value as it did when the resistor was connected directly across the bridge output terminal. When this condition occurs, the cable is exactly one-half wavelength long *electrically* at the test frequency. Before the clipping process begins, measure the cable and mark the calculated half-wavelength point with a band of tape around the cable. Proceed with caution as you approach the point on the cable where the meter begins to null deeply. In most cases, the calculated half wavelength and the measured half wavelength for new cable will be very close together.

The same procedure may be used to determine the electrical length of a line containing *any number* of multiples of half wavelengths (1/2, 2/2, 3/2, etc.). Once the length required for an electrical half wavelength, or multiple thereof, has been determined, the phase shift (delay) in degrees per foot may be calculated. The results are generally quite accurate *for the same batch of cable at a given frequency.* The design frequency for the array should be at approximately the center of the range of the operating frequencies. The author uses 7.2, 14.2, and 21.3 MHz as design-center frequencies for the 40-, 20-, and 15-meter bands, respectively. The performance of the array is usually satisfactory on the other frequencies within these bands.

Most radio amateurs and electronics technicians are aware that the *phase difference* between two ac signals may be determined by means of an oscilloscope. For phase-differential measurements, the oscilloscope must have identical vertical and horizontal deflection amplifiers. Also, the gain of each amplifier must be adjusted so that equal deflection is obtained at the vertical and horizontal crt deflection plates. Manufactured oscilloscopes with these features, particularly those capable of resolving the phase differences of rf signals in the region above 3.0 MHz, are quite expensive. The *vectorscope* is designed especially for phase-difference measurements and is used extensively in color-television laboratories and stations, but is priced beyond the reach of most radio amateurs. It is possible, however, to construct a small vectorscope at very

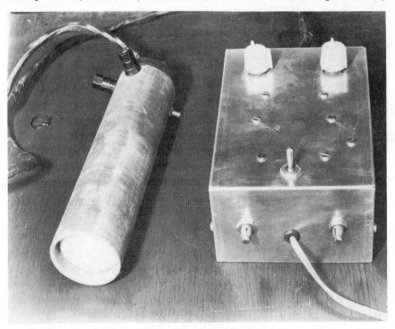

Fig. 3-8. Vectorscope.

little cost which can also be used for other purposes around the amateur radio station. Although such an instrument has its limitations, it will serve for most amateur purposes.

Fig. 3-8 shows one of several small vectorscopes built by the author and used at his station. The 2-inch crt is a type 902A which has an octal base and requires only 600 volts on the second anode. The circuit for the vectorscope is shown in Fig. 3-9. The voltage on anode No. 1 is 150 volts. The current requirements are very modest and the power supply utilizes a small 250-volt ac power transformer with a 20–25-mA secondary rating. The full-wave rectifier is connected in a voltage-doubler arrangement. The transformer filament winding should supply 6.3 volts ac at 1.0 ampere and should be insulated for at least 2000 volts. The 902A crt is listed as an obsolete type, but most surplus houses still have them in stock at a reasonable price.

The 600-volt power supply is contained inside a 3- × 4- × 6-inch aluminum box as shown in Fig. 3-8. The crt is enclosed inside a 9-inch section of bakelite tubing with an *inside* diameter of 2⅛ inches. The octal tube socket is placed on the tube base and the entire assembly is pushed inside the bakelite tubing. Two holes are drilled in the bakelite tubing at the socket end for mounting two binding posts. The crt is aligned so that the deflec-

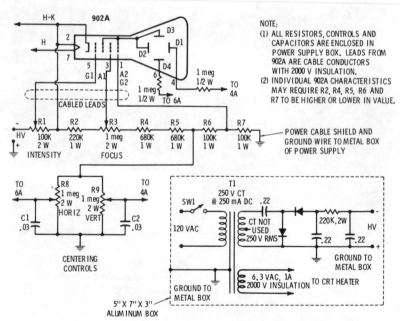

Fig. 3-9. Schematic diagram for vectorscope shown in Fig. 3-8.

tion-plate terminals of the tube socket are in contact with the binding-post terminals. The connections between the deflection-plate terminals and the binding posts are soldered resulting in not only a rigid mount for the crt but extremely short and direct connections to the deflection plates. As a result of the short deflection-terminal leads, the vectorscope is usable up to at least 30 MHz. This particular version has no horizontal or vertical amplifiers, and the rf signals are applied directly to the deflection plates. The rf voltages necessary to produce a display pattern of useful size should be between 50 and 100 volts peak-to-peak.

Make certain that the common ground connection between the crt and the power supply is in good condition. If the ground connection is broken, there will be a 600-volt potential difference between the crt and the aluminum case of the power supply. The control knobs shown in Fig. 3-8 are the vertical and horizontal centering controls. The two screwdriver adjustments on the side of the case are the focus and intensity controls.

The circuit shown in Fig. 3-10 is called a *quadrascope* (a form of vectorscope). The basic crt circuit is the same as that of the vectorscope previously described. The quadrascope, however, contains two tuned rf amplifiers and two matched grounded-grid rf input circuits. This instrument is designed for use at very low rf signal levels from the antenna system. The tuned circuits must be accurately aligned to the test frequency before using the instrument for phase measurements.

The vectorscope in Fig. 3-8 has a separate power supply and, since the interconnecting cable is about 5 feet long, the power transformer and the crt are relatively isolated from each other. In the quadrascope, however, the crt power supply is placed underneath the aluminum chassis. If an unshielded power transformer is used, the magnetic field of the transformer may cause a slight deflection of the crt electron beam. The most noticeable effect will be difficulty in obtaining a perfectly round sharp dot at the center of the crt screen. If the beam is affected by the transformer field, the beam may be elongated in the vertical or horizontal direction, or it may be difficult to focus the beam to produce a sharply defined dot. In the quadrascope, the power transformer is mounted on the inside rear chassis skirt. It was rotated during mounting to produce the least effect on spot shape and size. In all cases where the transformer and the crt are mounted on the same chassis, the use of a shielded power transformer is recommended. The Triad type R-104A or equivalent is a satisfactory power transformer.

The vectorscope shown in Fig. 3-8 is useful for the display of frequency ratios (Lissajous patterns), particularly at low fre-

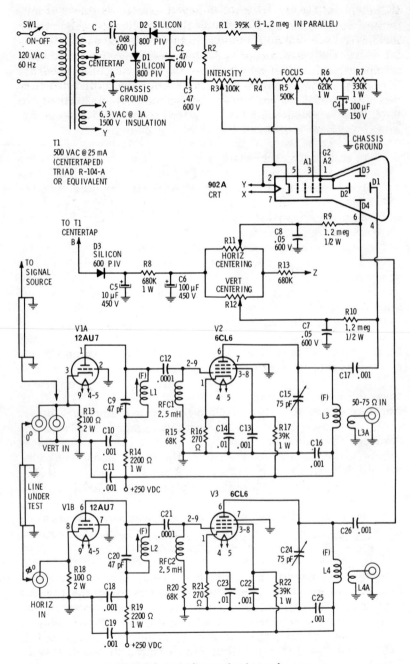

Fig. 3-10. Schematic diagram for the quadrascope.

quencies, and for checking the linearity of class-B linear rf amplifiers. For further information on these applications, the reader is referred to the *Radio Handbook* published by Howard W. Sams & Co., Inc. The quadrascope is designed especially for phase measurements, although it may be used to check amplifier linearity. Obviously, if the amplifiers are disconnected from the crt deflection plates and the signal under observation is applied directly to the plates, the quadrascope may be used to display Lissajous figures in the same manner as the simpler vectorscope.

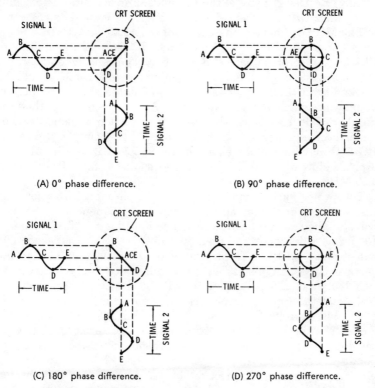

(A) 0° phase difference.

(B) 90° phase difference.

(C) 180° phase difference.

(D) 270° phase difference.

Fig. 3-11. Lissajous patterns for rf signals of the same frequency, with phase differences of 0°, 90°, 180°, and 270°.

When the Lissajous pattern of two signals with the same frequency is observed, the pattern may appear as one of the illustrations shown in Fig. 3-11. If the two frequencies are *exactly the same,* the pattern will remain fixed and the phase relationship of the two signals *with respect to each other* may be determined. If the frequencies of the two signals are slightly different, the pattern will change continuously, varying from one pattern to the next. In order to measure phase differentials with any degree of accuracy, a crystal-controlled reference signal is essential. It is advisable to obtain some practical experience in phase measurements by practicing with coaxial lines of various lengths.

The serious radio amateur should make up a series of 45°, 90°, 135°, and 180° cables fitted with standard PL-259 connectors at each end for test and calibration purposes. If the cables are made up for the highest hf band, say 15 or 20 meters, the sections may be coupled together (in series) to obtain almost any phase shift (delay) desired for any of the hf bands. For instance, a 45° cable at 7.1 MHz will be a 90° cable at 14.2 MHz. Likewise, a 45° cable at 14.2 MHz will have a phase shift of only 22.5° at 7.1 MHz. Short pieces of cable with 2°, 4°, 6°, 8°, and 10° phase shift, fitted with PL-259 connectors at each end, will be very useful when checking antenna system phasing. Each section of test cable should be marked with the number of degrees of phase shift and the *frequency* at which it was calibrated. Always remember that the number of degrees phase shift in a given section of cable depends upon the frequency at which it is used. For example, a piece of RG-8/U may have a phase shift of 10° at 7.1 MHz, but it will have a phase shift of 20° at 14.2 MHz and a phase shift of only 5° at 3.55 MHz.

The projection drawings in Figs. 3-12 through 3-18 show the resultant patterns that appear when the frequency of the sine wave applied to the horizontal deflection plates is first in phase (0° phase difference) and then 30°, 45°, 90°, 135°, 150°, and 180° out of phase with the frequency of the sine wave applied to the vertical deflection plates. At phase-difference angles other than 90°, 180°, and 270°, the pattern will be displayed as an ellipse. Note that the top of the patterns representing phase differences of 0°, 30°, and 45° leans toward the right. The 90° pattern is a circle. The top of the 120°, 135°, and 150° patterns leans toward the left. Also note the Y-axis intercept and Y-axis maximum points shown on the drawings. By the use of these points, *any* phase difference can be accurately determined. Use a calibrated scale on the oscilloscope and determine the points at which the pattern crosses the X and Y axes of the scale. The distance of the Y intercept from the center of the screen divided by the distance of the Y maximum from the center of the screen equals the *sine* of the phase-difference angle. Mathematically:

$$\text{sine } \theta = \frac{\text{Y intercept}}{\text{Y maximum}}$$

where,

θ is the phase-difference angle,

Y intercept is the distance from the center of the screen to the point where the ellipse crosses the Y axis,

Y maximum is the distance from the center of the screen to the highest point on the ellipse.

Detailed procedures for measuring phase differentials with the oscilloscope will be given in the next chapter.

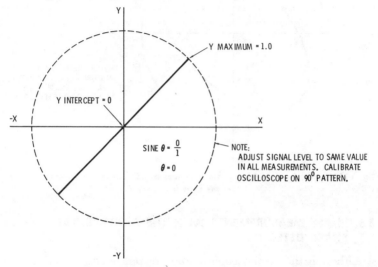

Fig. 3-12. Lissajous pattern for a phase difference of 0° between two sine-wave signals of the same frequency.

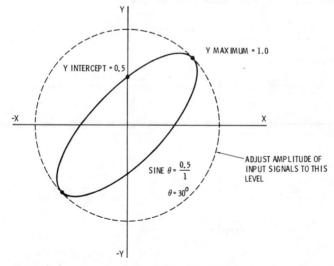

Fig. 3-13. Lissajous pattern for a phase difference of 30° between two sine-wave signals of the same frequency.

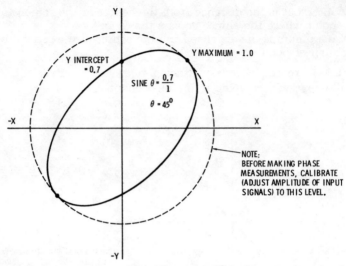

Fig. 3-14. Lissajous pattern for a phase difference of 45° between two sine-wave signals of the same frequency.

3-5. PHASE MEASUREMENTS WITH THE DIODE-BRIDGE PHASE DETECTOR

Although the oscilloscope method of phase-angle measurement is probably the most accurate, it is somewhat complex and the beginner must exercise care in interpreting the various Lissajous

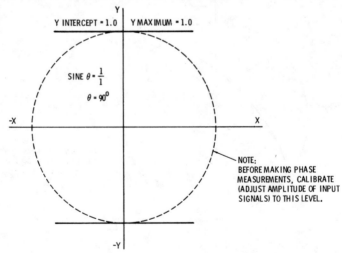

Fig. 3-15. Lissajous pattern for a phase difference of 90° between two sine-wave signals of the same frequency.

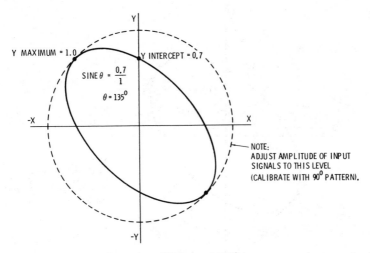

Fig. 3-16. Lissajous pattern for a phase difference of 135° between two sine-wave signals of the same frequency.

patterns obtained. The diode bridge shown in Fig. 3-19 has been used by the author, especially for quick checks on the system. The schematic diagram for the diode-bridge phase detector is shown in Fig. 3-20. The two diodes are type 1N270 or the equivalent. This phase detector functions in a manner similar to that of a phase detector in a color-television receiver or a phase discriminator in an fm receiver. Before using the diode-bridge phase de-

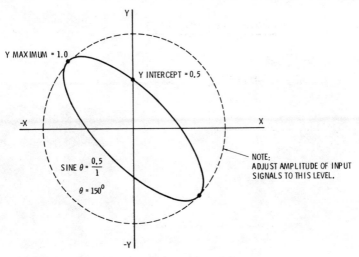

Fig. 3-17. Lissajous pattern for a phase difference of 150° between two sine-wave signals of the same frequency.

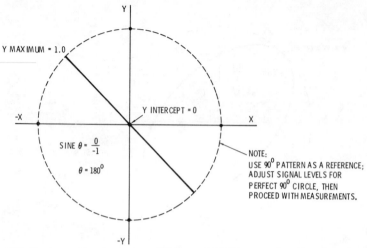

Fig. 3-18. Lissajous pattern for a phase difference of 180° between two sine-wave signals of the same frequency.

tector, the reader should study the following explanation and the calibration instructions carefully.

In the circuit shown in Fig. 3-20, the 0° reference rf signal is applied to chassis connectors J1 and J2. Coils L1 and L2 form a 4:1 step-up balun transformer that is independent of frequency and will operate from 3 to 30 MHz. Each coil consists of 15 turns, bifilar-wound and connected as shown on a 3½-inch long, ⅝-inch

Fig. 3-19. Diode-bridge phase detector.

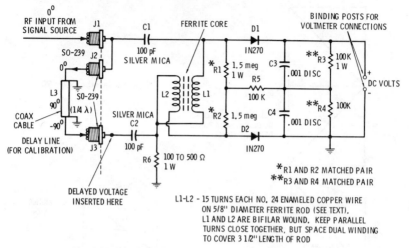

Fig. 3-20. Schematic diagram for diode-bridge phase detector.

diameter ferrite core from a broadcast radio "loopstick" antenna.
A small ferrite toroid core, such as the Amidon type FT-50, Mix-
72 core, can also be used. The number of turns and the coil connections will be the same as for the loopstick core. Due to the
high-Q of the coils, the rf voltage is about 15 volts peak between
the terminals at the diode inputs when the input from the signal
generator is about 3 volts peak-to-peak. The rf voltage at each
diode input should be the same when measured with a detector
probe and vtvm. If the voltages are not equal, there will be some
residual dc error voltage when the bridge is balanced. The rf
voltage across *one* of the coils will be *in phase* with the applied 0°
reference-signal voltage; the rf voltage across the *other* coil will
be *180° out of phase* with the applied 0° reference-signal voltage.
See the phasor diagram in Fig. 3-21. With only the 0° reference signal applied and no cable connected to chassis connectors J2 and
J3, the *positive* dc voltage across the binding posts will be about
5 volts read with a vtvm. When a coaxial cable exactly 90° (quarter-wavelength) long *at the 0° reference-signal frequency* is connected between J2 and J3, the voltmeter should read zero.

Referring to the phasor diagram in Fig. 3-21, the two rf voltages from the balun transformer are effectively applied to diodes
D1 and D2 in a push-pull or balanced relationship. The delayed
rf signal is fed to the transformer center tap and, therefore, is
effectively applied in phase to the two diodes.

Resistors R1 and R2 are 1-megohm, ½-watt, 1-percent resistors.
Resistors R3 and R4 are 100,000-ohm, ½-watt, 1-percent resistors.
The exact values are not so important but the two resistors in

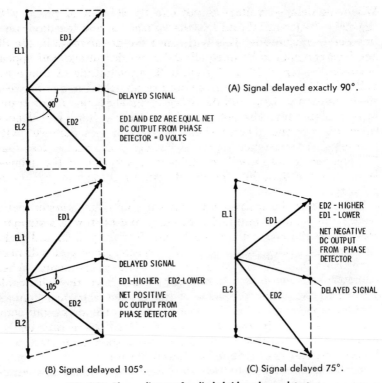

(A) Signal delayed exactly 90°.

ED1 AND ED2 ARE EQUAL NET
DC OUTPUT FROM PHASE
DETECTOR = 0 VOLTS

(B) Signal delayed 105°.

ED1-HIGHER ED2-LOWER

NET POSITIVE
DC OUTPUT FROM
PHASE DETECTOR

(C) Signal delayed 75°.

ED2 - HIGHER
ED1 - LOWER

NET NEGATIVE
DC OUTPUT
FROM PHASE
DETECTOR

Fig. 3-21. Phasor diagram for diode-bridge phasor detector.

each pair should be selected to have exactly the same value. The
resistor R5 is also a 100,000-ohm, ½-watt type, but its value is
not critical. Unless termination resistor R6 is used, no zero indication will be obtained when the delayed signal is injected
through C2 to the transformer center tap. Try values of 100 to
500 ohms for R6. Use the *lowest* resistance value that will still
permit a stable zero dc voltage indication. If the instrument case
is "hot" with rf voltage, the zero voltage setting will tend to be
erratic. If the zero indication fluctuates, connect the phase-detec-
tor and signal-generator cases together and to an earth ground.
When using the instruments to adjust the antenna phasing, always connect the cases to the array ground or radial system.

When the *delayed voltage* applied to the transformer center
tap is *exactly 90° out of phase* with the coil voltages at D1 and
D2, the voltages across the two diodes are equal in magnitude.
Then, the dc currents in the diode load resistors R3 and R4 are
equal and opposite. Under these conditions, a net *zero* dc voltage
will appear between the top of R3 (output terminal) and ground.

When the delayed voltage is *not* exactly 90° out of phase with the coil voltages at D1 and D2, the voltages across the diodes are unequal in magnitude. This will cause the dc currents in the diode load resistors to be unequal and a net dc voltage will appear between the top of R3 and ground. The magnitude of this *error* voltage will depend upon the amount of phase difference between the 0° reference signal and the delayed signal applied to the transformer center tap. The polarity of the error voltage will depend upon whether the delayed signal is more or less than 90° with respect to the reference signal. If the delayed signal is *less* than 90°, the net dc output voltage polarity is *negative*. If the delayed signal is *more* than 90°, the net dc output voltage is *positive* in polarity.

To calibrate the instrument, make up an *exact quarter-wavelength* (90°) line as outlined previously and fit it with a standard PL-259 coaxial connector at each end. In fact, you should make up *two* such cables, *identical in all respects*, since you will need a second cable for actual antenna-phasing adjustments. When calibrating or using the instrument, make sure that the cable plugs and the chassis receptacles make good electrical contact. If the connector pins are dirty or tarnished, the dc output indications, particularly the zero indication, will be erratic or incorrect. The connector pins should be polished with steel wool before calibrating or using the phase detector.

When you are certain that your delay test cables are cut to exactly 90°, connect a vtvm or equivalent transistorized voltmeter to the dc output terminals and apply the 0° reference signal only to chassis connector J1. At this time do not connect the delay cable. Set the voltmeter to the +5-volt dc range and adjust the rf signal generator output level to give exactly *full-scale* indication on the voltmeter. When making checks with the delay cable connected, the full-scale voltage adjustment may be made by removing the cable connector from connector J2 only. It is not necessary to remove the delay cable completely from the circuit. Now, with the delay cable connected to J2 and J3, the voltmeter should read very close to zero. Adjust the voltmeter zero control for a zero voltage reading. Do not change the signal level from the generator.

The delay cable is removed from J2 and J3 after calibration and the phase detector is ready for use to check or adjust the antenna system. Although the procedure given here for antenna phasing adjustments is brief, the essential points are covered. For more detailed procedures, refer to the later chapters, particularly Section 5-5. Always check the instrument calibration before making actual adjustments of the antenna phasing system. If you obtain readings that are unusual or do not appear to be logical during actual phasing adjustments of the array, always check plug connections and the instrument calibration. The measurement of phase in an antenna system is, at best, a critical process and it is necessary to exercise extreme care in making connections to the antenna elements and the radial system or ground.

To make phase measurements in the antenna system, you will need *two* identical quarter-wavelength (90°) coaxial cables fitted with PL-259 plugs. For array measurements, connector J2 is not used. The main transmission line is disconnected from the transmitter and the generator signal is applied to the line at the transmitter end. One 90° test cable is used to pick up the rf signal at the 0° reference antenna element. The 0° reference antenna signal is applied to J1. The other 90° test cable is connected to J3 and picks up the delayed rf signal at the other antenna element. Before inserting the delayed signal, however, check with only the 0° antenna signal connector and adjust the voltmeter for full scale reading as described above. Next, connect the delayed signal cable to J3 and observe the voltmeter indication. If the system is properly phased (90° lag at the delayed element) the voltmeter should indicate zero, just as it did with the 90° delay test cable connected between J2 and J3. If a positive or negative dc voltage is indicated, the antenna system is not properly phased and the phasing must be adjusted until the voltmeter reads zero. The exact steps to be taken will vary with the type of array. The proper methods for adjusting several different antenna systems are covered later in the book.

Do not connect the coaxial test cables directly to the antenna elements. Unless some isolation is provided, the test cables will

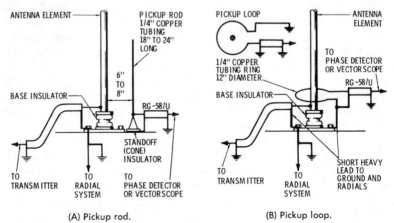

(A) Pickup rod.   (B) Pickup loop.

Fig. 3-22. Pickup elements for antenna phase measurements.

detune the elements and probably upset the impedance match. Even if you are able to obtain readings that appear to be correct, the system will go out of adjustment the instant the test cables are removed. Practical methods of extracting the 0° and 90° sampling signals are covered in detail in later chapters. The usual technique is to use either short pickup rods parallel with the antenna element (Fig. 3-22A) or large copper-tubing ring coils around the low-voltage portion of the antenna element (Fig. 3-22B). The author has occasionally used a 4.7-megohm isolation resistor in series with the inner conductor of the coaxial test cable at the antenna for signal pickup. The pickup loop appears to be the best method. The loop, which consists of one turn of ¼-inch copper tubing, is about 12 inches in diameter and is placed around the antenna element near the base. One end of the loop is grounded to the radial or ground system; the other end is connected to the inner conductor of the coaxial test cable. The pickup loops may be permanently installed for occasional checks of the array phasing. For horizontal arrays, including yagis, quads, etc., the pickup rods may be suspended a few inches below the center of the element by means of nylon cords.

3-6. MULTIPLE-PURPOSE RF TEST INSTRUMENT

The test instrument shown in Fig. 3-23 is designed to make maximum use of a 0–200-μA dc microammeter. Fig. 3-24 shows the schematic for this instrument which performs as a field-strength meter, a phase meter, and a swr meter or *monimatch* reflectometer. Any of these three functions may be selected by SW1 which is a single-pole, three-position rotary switch. Switch SW2 is a single-pole double-throw type which selects either *forward* or *reverse* line signal indication when switch SW1 is placed in the monimatch position. When SW1 is in the field-strength-meter position, the signal from the high-Q resonant circuit is rectified by diode D1 and the resultant dc voltage is applied to the microammeter. In the phase-meter position, the meter is used as a visual indicator for the diode-bridge phase detector described previously.

The monimatch sampling element consists of three thin copper strips cemented and bolted to a perforated circuit board or fiberglass plate and arranged as shown in Fig. 3-25. The center strip is first cemented and bolted in place and soldered to the inner-conductor terminals of the two SO-239 coaxial connectors. Before cementing and bolting the side strips in place, however, they are held with small alligator clips and the air-gap spacing is adjusted. The output (antenna) terminal of the unit should be terminated

Fig. 3-23. Multiple-purpose rf test instrumtnt.

in either a 50- or 72-ohm carbon resistor (several resistors paralleled to produce the correct resistance with about 5 watts of dissipation capability) or a well-matched 50- or 72-ohm transmission line. An rf signal is applied to the input terminal and switch SW2 is placed in the *forward* position. Gradually increase the level of the RF signal until an indication is obtained on the 0–200-μA dc microammeter. Adjust the spacing between side strip ST1 and the center strip for maximum indication on the microammeter. If the pointer goes off-scale during this adjustment, reduce the sensitivity of the meter with the sensitivity control. Now, place switch SW2 in the *reverse* position. The meter reading should drop sharply and may go to zero with resistance termination. With transmission line and antenna termination, the

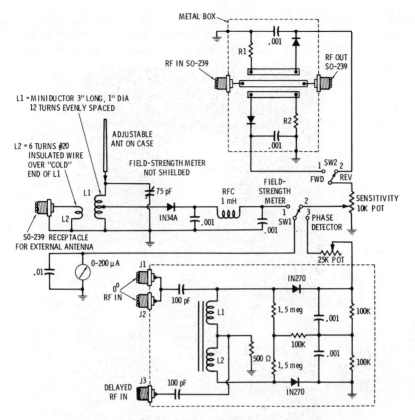

Fig. 3-24. Schematic diagram for the multiple-purpose rf test instrument.

meter will not read zero if standing waves are present on the line. Dress terminating resistor R2 at right angles to the center strip to prevent inductive coupling effects. If the meter does not read very near zero after dressing R2, try slightly higher or lower values for resistors R1 and R2.

The next step is to reverse the input and output connections to the sampling element. Feed the rf signal to the output terminal and connect the 50- or 72-ohm load to the input terminal. The signal is now being sent through the sampling unit in a direction opposite to that of normal use. Place switch SW2 in *reverse* position and adjust the spacing between side strip ST2 and the center strip for maximum reading on the 0–200-μA microammeter. Observe the same precautions as outlined above. Switch SW2 to the *forward* position and dress R1 for minimum meter reading. When optimum adjustments for maximum and minimum meter readings have been obtained, seal the strips in position with a drop or two of epoxy cement. After the cement has hardened, drill small holes through the strips and perforated board as shown in Fig. 3-25, and bolt the strips and board together. The plated brass machine screws should be used instead of steel.

When connected as shown, the unit will pass a full kilowatt of rf power on cw and 2500 watts pep on ssb radiotelephone without arcing or breakdown between the strips. If desired, the unit may be left permanently in series with the transmission line.

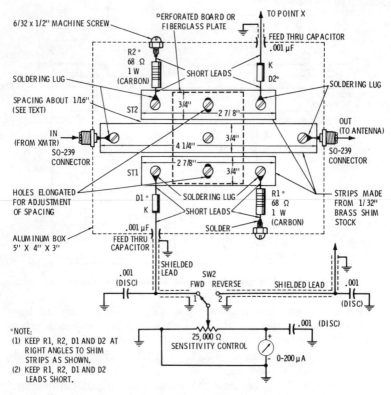

Fig. 3-25. Construction details for monimatch.

The field-strength-meter section uses a high-Q tuned circuit. To prevent loading L1, diode D1 is connected to the center of the coil. Coil L1 is a B & W Type 3013 air-wound Miniductor and consists of 12 turns of No. 16 wire, 1 inch in diameter and 3 inches long. Coil L2 consists of 6 turns of insulated hook-up wire wound over the "cold" end of L1. The adjustable antenna connected to the top of L1 is the type used with small portable transistor radio receivers. A coaxial fitting is provided for connecting a coaxial cable with a pickup coil or external antenna. Detailed instruction on the use of a field-strength meter for relative field-strength measurements will be given later.

If the meter reading is reversed in any of the switch SW1 positions, check the polarity of the diode in that part of the instrument circuit being used. When buying bulk or "bargain-package" diodes, always check the front-to-back resistance ratio and polarity with an ohmmeter before wiring them into the circuit. Some of these "surplus" diodes may be incorrectly marked.

SELF-EXAMINATION

Here is a chance to see how much you have learned about antenna measurements and test equipment. These exercises are for self-testing only. Answer true or false.

1. Frequency stability in a test-signal generator is not important.

2. A resistance impedance bridge will indicate reactance at an antenna feedpoint or in a transmission line.

3. The R-X bridge can be calibrated to indicate both resistance and reactance in a transmission line or antenna system and will also indicate the sign of the net reactance.

4. In a single-element antenna system, the phase of the rf current at the antenna feedpoint is of no importance.

5. In multielement radiating systems, the connecting leads to relays, switches, and similar control devices may cause undesired phase shift at the antenna feedpoints.

6. When making phase measurements in an antenna system, all phase references are with respect to the phase of the signal at the transmitter output terminal.

7. When two rf signals of the same frequency are applied to the vertical and horizontal plates of the quadrascope and the display is a perfect circle, the two signals are 90° out of phase with each other.

8. Refer to Question 7 above. If the vertical plate signal is taken from a reference-phase antenna through 100 feet of RG-58/U coaxial cable and the horizontal plate signal is taken from another antenna through 60 feet of RG-58/U cable, then a perfect circle displayed on the quadrascope indicates that the current at the second antenna feedpoint is 90 degrees out of phase with the current at the reference antenna feedpoint.

9. The diode-bridge phase detector is useful for measuring phase differentials in the 90-degree region. The quadrascope is more accurate for all phase measurements.

10. When making phase measurements and adjustments on an antenna system, the test cables may be connected directly to the antenna-element feedpoints through a small-value capacitance. Because of the low impedance at the feedpoints, the test line connection will not affect the phase of the rf currents in the antenna elements.

CHAPTER 4

Impedance Measurements and Matching Procedures

The process of transforming the resistive component of an antenna or feedline impedance to a desired value of load resistance is generally called *impedance matching*. Actually, no matching of any kind takes place. The impedance is simply made resistive by canceling the reactive component, and then it is transformed to a value which the final rf amplifier "sees" as a load resistance equal to its own internal impedance. All of the various matching adjustments in the transfer circuitry from the antenna back to the final rf amplifier plate are made for only one purpose—to make the rf amplifier plate "think" that it is working into a *pure-resistance* load. As an example, we may have a linear rf amplifier with a plate impedance of 5000 ohms.

We can prove that in order to produce maximum power in a load, it is necessary that the load resistance equal the internal resistance of the source. Therefore, the tuned tank circuit across the plate circuit of the tube must be designed to appear as a 5000-ohm *resistive* load to the tube plate. Note the use of the word "resistive" instead of "resistance." Loads in rf circuits are rarely resistance loads but are made to appear as such by proper design and adjustment of the components involved. In the case of our hypothetical linear rf amplifier, we must design the plate tank circuit in such a manner that the 5000-ohm value required to match the tube plate impedance is *stepped down* to some value that will match the impedance of a standard coaxial transmission line, usually 50 or 75 ohms. If we connect a 50- or 75-ohm resistive load having suitable power-dissipation capabilities across the low-impedance output terminals of the tank circuit, the low resistance value (50 or 75 ohms) will be *reflected* across the tank circuit and will appear to the tube plate as the proper 5000-ohm load. If some other value of load resistance (for example, 500 ohms) is connected across the low-impedance output terminals, the tank circuit will not present the proper load value to the plate of the tube. This will result in a serious *mismatch* and lack of power transfer from the tube plate to the tank circuit.

When an incorrect load is placed across the tank-circuit output terminals, such as by improper adjustment of the antenna system, reactance (either X_L or X_C) will be reflected back into the plate tank circuit of the rf amplifier. This will cause the tank circuit to be detuned and, if nothing else, can be a nuisance, particularly in switched-direction, phased arrays where the swr varies with different switch positions. In some phased arrays, especially the close-spaced, compact types, the swr on the line may vary by a factor of 2:1 when switching from end-fire to broadside operation. This much swr *change* will generally require readjustment of the amplifier tank circuit each time the direction of transmission is changed. In a fixed-direction installation, or when using a horizontally polarized rotary beam, this effect will not be noticed since the reflected reactance would be tuned out by the initial adjustment of the tank circuit. Obviously, the antenna feedpoint must be matched to the transmission line; otherwise, the mismatch will be reflected all the way back up to the plate of the rf amplifier as previously described.

Impedance mismatch conditions are much more apparent when the final rf amplifier is working into a switchable vertical array. This is one of the reasons for frequent reference to vertical arrays during our discussions. Now, let us consider the various methods of impedance-matching transmission lines to the antenna feedpoint and the proper measurement and adjustment procedures.

4-1. THE L-NETWORK MATCHING SYSTEM

The L-network, illustrated in Fig. 4-1, is by far the most commonly used circuit for matching a low-impedance source to either a lower- or higher-impedance load. It is ideal for matching either 50- or 75-ohm coaxial lines to the base of vertical antennas where the impedance must be stepped down, or to single-wire antennas where the impedance must be stepped up in value. The circuit shown in Fig. 4-1A is used when the antenna feedpoint impedance is higher than the line impedance. The circuit in Fig. 4-1B is used when the antenna feedpoint impedance is lower than the line im-

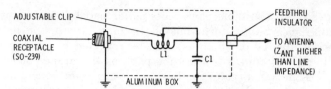

(A) Used when antenna feedpoint impedance is higher than feedline impedance.

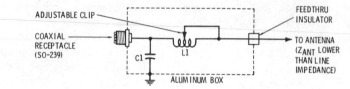

(B) Used when antenna feedpoint impedance is lower than feedline impedance.

Fig. 4-1. L-network matching system.

pedance. These circuit arrangements may be used at the transmitter to feed a single-wire antenna of random length, at the antenna feedpoint, or inserted into the transmission-line network where it is desired to step-up or step-down the impedance value. The L and C values will depend upon the impedance value at the point of insertion and the frequency involved.

A practical example of L-network application is illustrated in Figs. 4-2 and 4-3. This antenna coupler-matching unit is used by

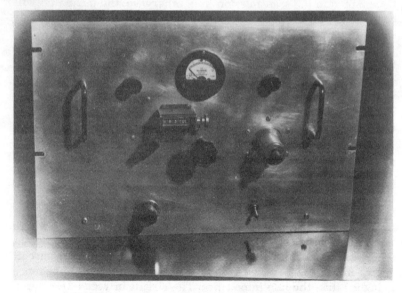

Fig. 4-2. Front-panel view of the L-network antenna tuner.

Fig. 4-3. Rear view of the L-network antenna tuner.

the author to feed rf power from the 50-ohm final amplifier terminal of his station to various types of single-wire antennas. The schematic diagram is shown in Fig. 4-4. Inductance L1 is a tapped coil that has a total of 36 turns of No. 12 enameled copper wire wound in 4 segments of 9 turns each on a 2-inch form. Each 9-turn segment is close wound and the 4 segments are connected in series. The start and finish of coil L1 and the junction of each segment are connected to the terminals of a single-pole, 6-position heavy-duty rotary switch, as shown in Fig. 4-3. Coil L2 is a surplus roller coil with an inductance of about 10 microhenries. Capacitor C1 is a 250-pF mica type rated at 2500 volts. Capacitor C2 is a Jennings vacuum-variable type with a capacitance range of 10 to 750 pF. Switch SW3 consists of a flexible copper-braid lead attached to the "hot" side of capacitor C2 on one end and terminated in an alligator clip on the other. The alligator clip is attached to the input or output terminal of coil L2, depending upon whether the antenna feedpoint impedance is higher or lower than the coaxial-line impedance. For end-feeding a high-impedance, long-wire antenna, the switch position shown in the schematic is most generally used.

In order to indicate tuner resonance, an swr bridge or monimatch circuit must be inserted in series with the low-impedance

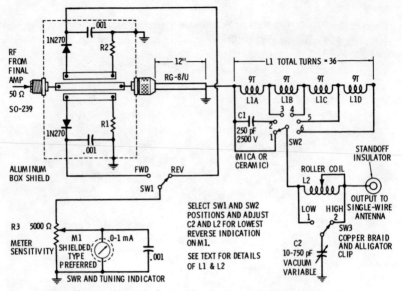

Fig. 4-4. Schematic diagram for the L-network antenna tuner.

line from the transmitter. All tuning adjustments are made to obtain proper loading of the final rf amplifier plate circuit consistent with the lowest swr on the low-impedance line. The procedure begins with the r-f amplifier adjusted for an output impedance of approximately 50 ohms and a single-wire antenna connected to the tuner output terminal. Keep the amplifier rf output low and adjust the tuner sensitivity control so that about half-scale indication is obtained on the 0–1-mA meter with switch SW1 in the *forward* position. Adjust coil switch SW2, the roller coil, and capacitor C2 for maximum reading on the meter. As the tuning progresses, reduce the sensitivity of the meter to keep the needle within the scale limits. After resonance has been established, place switch SW1 in the *reflected* or *reverse* position and adjust the tuner controls for minimum indication on the meter. If the sensitivity control is adjusted for exactly full-scale deflection on the meter in the forward position and the meter indicates exactly zero when switch SW1 is placed in the reverse position, the swr on the low-impedance line will be 1:1 ratio—a perfect match. Using the transmitter loading controls, load the amplifier to the desired power input level.

The L-network in the low-pass configuration will aid in reducing harmonic radiation. However, because the impedance of the antenna at a particular harmonic may be much different from the impedance at the fundamental, the L-network may not offer much attenuation to harmonic radiation under certain conditions. Therefore, a *low-pass filter* should always be inserted in series with the low-impedance coaxial line as close as possible to the transmitter. It is imperative that the antenna tuning and amplifier loading be carried out with a low swr on the line. Some low-pass filters such as those made by Johnson and Collins are easily damaged when operated in a line with high swr.

The L-network shown in Fig. 4-5 is being used to match the impedance of a vertical antenna feedpoint to a 50- or 75-ohm

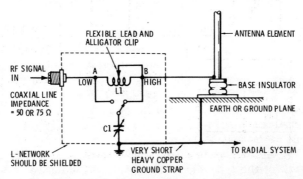

Fig. 4-5. L-network used to match vertical antenna to line.

coaxial line. The following procedure is used for measuring the antenna base impedance in order to determine which type of L-network configuration will be required. (1) Connect the R-X bridge (Section 3-3) output terminal to the antenna and ground system as shown in Fig. 4-6. Make certain that all instruments

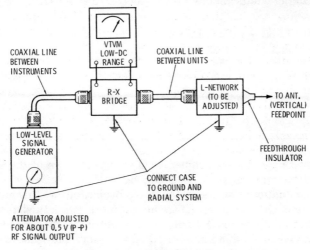

Fig. 4-6. Hookup of test equipment to adjust L-network.

are properly grounded. Feed a low-level (about 0.5 volt peak-to-peak) rf signal of the correct frequency to the bridge input terminal. Set the reactance (X) dial to 0 and rotate the resistance (R) dial for a null in the indicator (vtvm) reading. If a null is obtained but the meter does not indicate near zero volts, rotate the X dial for the deepest possible null. The X and R dials may require slight readjustments to obtain the lowest possible meter reading in the null depth. Note whether the X dial indicates that more capacitance (less X_C) or less capacitance (more X_C) is required to establish the deepest null.

It is *not* likely that the base impedance of the quarter-wavelength vertical antenna will measure 50 or 75 ohms—a match for a standard coaxial line. If the antenna impedance measures higher than 36 ohms when both dials are adjusted for the greatest null, it is quite likely that the radial or ground system is inefficient and steps should be taken to improve it before making further measurements. Most quarter-wavelength vertical antennas, when cut to the formula, exhibit inductive reactance and require added X_C in order to bring them to resonance. In designing an L-network of the type shown in Fig. 4-5, it is desirable to begin with a variable capacitor for C1. The author uses an old 4-section broadcast–receiver-type tuning capacitor with all four fixed-plate sections connected in parallel. The total full-mesh capacitance is about 1000 pF. At half scale the total capacitance is about 500 pF, and at one-fourth scale it is about 250 pF. Exact calibration of the capacitor is not necessary, since it is used only to estimate the *fixed-capacitance* value required in the final network. The construction data for coil L1 is given in Table 4-1.

Table 4-1. Construction Data for Coil L1 in Fig. 4-5

Band (Meters)	Diameter (Inches)	Turns	Spacing (Inches)	Conductor
40	3	12	1/8	1/4" Copper Tubing
20	2 1/4	7 3/4	3/16	3/16" Copper Tubing
15	2 1/4	5	1/4	3/16" Copper Tubing
10	2	3	1/4	3/16" Copper Tubing

Connect the variable capacitor C1 from the base of the antenna to ground (radial system) as shown in Fig. 4-7. Connect an *adjustable* inductor between the R-X bridge output terminal and the base of the antenna. Set C1 at full-mesh and adjust the R dial to either 53 ohms or 73 ohms, depending upon the impedance of the coaxial cable, and adjust the X dial to zero. Now, move the tap connection along the coil L1 toward the antenna, one turn at a time, while observing the bridge indicating meter. If no null is obtained, set capacitor C1 at about half mesh and repeat the procedure. If the antenna base impedance is higher than the line impedance, a null should be obtained at some adjustment of C1 and L1. *Do not use the bridge dials to produce a null.* The bridge dial should be set as described previously and then left alone during the matching adjustments. If a null appears, carefully adjust both C1 and L1 for the greatest null depth. The bridge indicator should read very close to zero when the correct adjustments have been made. If no null is obtained with the circuit of Fig. 4-5, disconnect capacitor C1 from the antenna and connect it to the bridge end of coil L1 and ground. Set C1 at full mesh and move the antenna tap connection along the coil toward the bridge, one turn at a time. Try different settings of capacitor C1 and taps along coil L1 until a null is obtained. Adjust C1 and L1 for the greatest null depth.

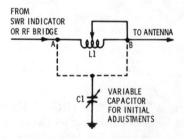

Fig. 4-7. Connection of variable capacitor for preliminary L-network adjustments.

Now, remove capacitor C1 from the circuit and either measure or estimate the amount of capacitance required. Select a fixed mica capacitor of the same approximate capacitance as the estimated value of C1 and connect it in the antenna circuit in place of C1. For high-power operation (600 to 1000 watts rms carrier) the fixed capacitor should have a working dc voltage rating of at least 2500 volts. For low-power operation, a capacitor rated at 1000 dc working volts will be satisfactory. Ceramic capacitors, such as the Centralab 850S series, may be used in place of the mica units. With the fixed capacitor in place, adjust the tap position on coil L1 for the greatest null. Once the correct L1 tap is located, solder the connection and cut off the unused portion of the coil. Coil L1 should be wound of No. 12 copper wire or ⅛-inch copper tubing. The turns should be rigidly supported or wound on a form to prevent inductance change. The coil and capacitor may be protected from the weather by placing them inside a plastic freezer-storage container.

4-2. THE PI-NETWORK MATCHING SYSTEM

While the L-network is the simplest and easiest-to-use imped-ance-matching device, the pi network is sometimes used to match impedances that are low in value and nearly equal. The pi net-work is almost universally used in the plate tank circuits of mod-ern transmitters and is often used to match the 50-ohm output impedance at the transmitter to other coaxial-line impedance val-ues. The matching circuit shown in Fig. 4-8A is useful for match-

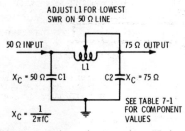

(A) Pi network used to match a 75-ohm line to a 50-ohm output.

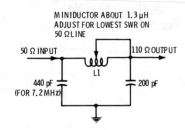

(B) Pi network used to match the rf output of a 50-ohm exciter to the 110-ohm input (cathode) of a grounded-grid linear amplifier.

Fig. 4-8. Pi-network circuits.

ing a 75-ohm line (RG-11/U) to a transmitter output impedance of 50 ohms. If the pi-network is properly adjusted, the swr on the line between the transmitter output and the network input will be 1:1. The pi network is useful for matching a single coaxial line (RG-8/U or RG-11/U) to two or more paralleled-input coaxial lines in power-divider systems. Another use for the pi network, other than as an antenna-matching circuit, is to match the output of a 50-ohm single-sideband exciter to the input circuit of a grounded-grid linear amplifier (Fig. 4-8B). The network is ad-justed for minimum swr on the interconnecting cable from the exciter rf output to the input of the linear amplifier in exactly the same manner as matching a transmission line to an antenna.

4-3. THE GAMMA MATCHING SYSTEM

The majority of amateur antennas are fed with coaxial cable. One reason is that interference-reduction problems, particularly tvi, are simplified when the antenna is fed from a shielded cable. Most of the modern amateur-station accessories, such as low-pass filters, swr and power-output meters, etc., are designed for inser-tion in coaxial lines.

The *gamma match,* designed for 50- or 75-ohm coaxial cable use with a horizontally polarized antenna, is illustrated in Fig. 4-9A. In this version, only *one* resonating capacitor is used. A modified version of the gamma match, in which *two* resonating capacitors are used, is sometimes referred to as the *"omega" match.* The matching functions of the two versions are the same; however, the adjustment procedures are slightly different. While the gamma and omega matching systems are generally used in con-

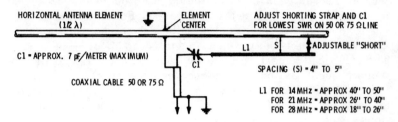

(A) Gamma feed for horizontally polarized antenna.

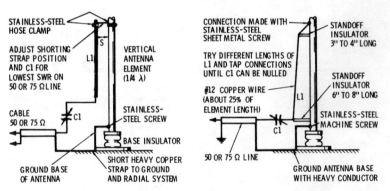

(B) Gamma feed for vertical antenna.

(C) Shunt feed for towers and vertical elements in the 80- and 160-meter bands.

Fig. 4-9. Gamma matching system.

nection with horizontally polarized, multielement yagi beam antennas, they are also applicable to low-impedance vertical antennas, as shown in Fig. 4-9B. A gamma-feed system called *shunt feed* (Fig. 4-9C) is widely used to feed rf power to towers and other vertical elements on the 75- and 160-meter amateur bands. For dimensions and other design data applicable to a specific antenna system or band of frequencies, the reader is referred to one of the many good radio-amateur antenna handbooks on the market. The adjustments and measurements on gamma and omega matching networks will be covered here.

The gamma match circuit, as shown in Fig. 4-10, is, in effect, a shorted-line section less than one-quarter wavelength long. As discussed previously, the reactance presented at the open end will be inductive. The purpose of capacitor C1 is to introduce an opposite-sign (X_C) reactance into the circuit so that a purely resistive feedpoint termination is presented to the coaxial line. The value of the feedpoint impedance is determined by the length of the gamma rod and its spacing with respect to the antenna conductor. The *antenna element should be grounded at the point opposite the open end of the gamma rod.* For a horizontal Yagi beam antenna, the center point of the driven element should be grounded to the boom through a metal strap about ¾ inch wide. The electrical contact is made by securing the strap to the antenna element and boom with *stainless-steel* sheet metal screws. In a vertical antenna system, make certain that the base of the antenna is grounded to the radial system. Unless these precautions are observed, the ground connections may tend to *float* causing considerable difficulty in obtaining accurate and stable adjust-

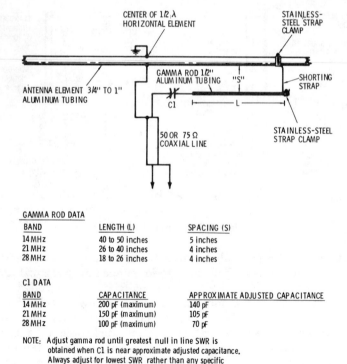

GAMMA ROD DATA

BAND	LENGTH (L)	SPACING (S)
14 MHz	40 to 50 inches	5 inches
21 MHz	26 to 40 inches	4 inches
28 MHz	18 to 26 inches	4 inches

C1 DATA

BAND	CAPACITANCE	APPROXIMATE ADJUSTED CAPACITANCE
14 MHz	200 pF (maximum)	140 pF
21 MHz	150 pF (maximum)	105 pF
28 MHz	100 pF (maximum)	70 pF

NOTE: Adjust gamma rod until greatest null in line SWR is obtained when C1 is near approximate adjusted capacitance. Always adjust for lowest SWR rather than any specific value of C1 or gamma-rod length.

Fig. 4-10. Gamma-rod dimensions for the 14-MHz, 21-MHz, and 28-MHz amateur bands.

ments. Also, antenna systems with floating or poor grounds tend to be noisy when used for reception.

The adjustment procedure, using the R-X bridge, is relatively simple. Make a rough adjustment of the gamma-rod length to the approximate dimensions shown in Fig. 4-10 and set capacitor C1 at about half mesh. The R-X bridge resistance dial should be set to a value equal to the impedance of the coaxial cable being used. The reactance dial should be set at zero. The test signal may be applied through either the coaxial cable that is used to feed the system or a separate coaxial line. With the test equipment set up as shown in Fig. 4-11, adjust capacitor C1 for the best null on the

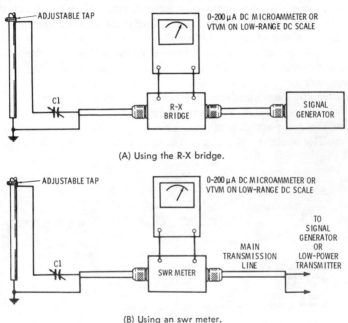

(A) Using the R-X bridge.

(B) Using an swr meter.

Fig. 4-11. Test set-up for making gamma-rod adjustments.

bridge indicator. The chances are that a null will be obtained but that it will be shallow indicating reactance is present at the gamma-match feedpoint. Rotate the reactance dial on the bridge until the greatest null depth is obtained. Note whether it was necessary to increase or decrease the capacitance of the bridge variable capacitor in order to produce the null. Now, if required, adjust the position of the shorting bar on the gamma rod and antenna element two or three inches in either direction and re-adjust the reactance dial for the null. If the shorting bar is being moved in the correct direction, the null will appear on the reactance dial closer to the zero calibration point. Continue to adjust the position of the shorting bar until the null occurs at exactly zero on the reactance dial. Readjust capacitor C1 for the best null.

The line swr at the transmitter output should be checked with a reflectometer or monimatch. If the swr is higher than about 1.4:1 *at the test frequency,* recheck the matching adjustments at the antenna. If the null obtained at the antenna is deep (almost zero indicator reading), measure the impedance of the coaxial cable at the transmitter end. If the null depth does not equal that obtained at the antenna feedpoint, or if reactance is indicated in the line by the null appearing at some point on the reactance dial other than zero, the coaxial cable may be defective or of poor quality. Old surplus cables, particularly those with dents, are a frequent source of swr problems.

Another source of swr problems is the common practice of suspending coaxial cables in the manner illustrated in Fig. 4-12A. The larger cables, such as RG-8/U and RG-11/U, have considerable weight per hundred-foot length. If long sections of these cables are suspended in the manner shown, the summer sun will cause the temperature of the dielectric to rise to the softening point. The weight will cause the cable to stretch, changing its

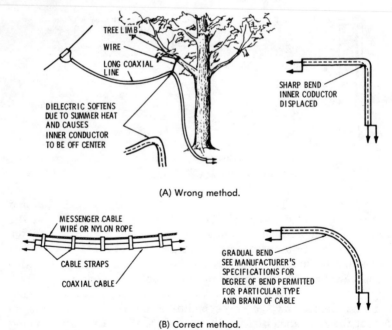

(A) Wrong method.

(B) Correct method.

Fig. 4-12. Suspending coaxial cables.

physical dimensions and, as a result, its electrical characteristics. In commercial installations, long runs of coaxial cable are supported on *messenger cables* as shown in Fig. 4-12B. For a good example, observe the manner in which telephone cables are supported between poles. Also, do not make sharp bends in coaxial cables or hang large loops of unsupported cable. When the dielectric of large unsupported loops of cable is softened by summer heat, the inner conductor may be displaced off-center. This will result in an electrical "bump" in the line and in an increased swr as measured at the transmitter end. Transmission-line cables should *not* be trimmed to obtain a low swr indication at the transmitter end. By this method an apparent 1:1 swr indication can be obtained. However, the amateur who practices this deception is only kidding himself—the swr is still the same on the line even though he has the low swr indication at the measurement point. These false indications will be discussed in more detail later.

4-4. THE OMEGA MATCHING SYSTEM

The modified gamma match illustrated in Fig. 4-13 is generally referred to in the amateur literature as the *omega* match. When using the gamma match, it is necessary to change the gamma-rod length by adjusting the shorting bar when the inherent inductive

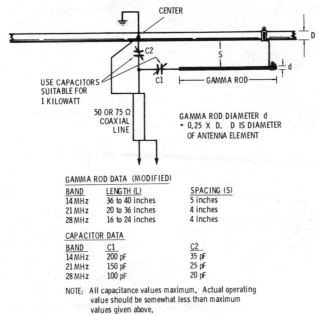

GAMMA ROD DATA (MODIFIED)

BAND	LENGTH (L)	SPACING (S)
14 MHz	36 to 40 inches	5 inches
21 MHz	20 to 36 inches	4 inches
28 MHz	16 to 24 inches	4 inches

CAPACITOR DATA

BAND	C1	C2
14 MHz	200 pF	35 pF
21 MHz	150 pF	25 pF
28 MHz	100 pF	20 pF

NOTE: All capacitance values maximum. Actual operating value should be somewhat less than maximum values given above.

Fig. 4-13. Modified gamma match.

reactance cannot be tuned out with capacitor C1. It is sometimes inconvenient, as well as dangerous, to adjust the gamma-rod length while working on a high tower or pole. In order to eliminate these difficult adjustments, the arrangement shown in Fig. 4-13 was developed. Instead of making physical adjustments of the shorting bar, the same electrical effect may be produced by fixing the position of the shorting bar and adjusting the added capacitor, C2. As a result of these changes, the correct adjustment procedure for the omega match is slightly different from that of the regular gamma match.

The R-X bridge and the signal source are connected in the same manner as previously described for the gamma-match adjustments. Gamma capacitor C1 and omega capacitor C2 are set at about half mesh. The bridge resistance dial is set to the impedance value of the coaxial cable being used. The bridge reactance dial is set at zero. Now, *alternately* adjust capacitors C1 and C2 for greatest null depth. Usually, the best procedure is to adjust C1 for the null and adjust C2 for depth. After adjusting C2, it will be necessary to readjust C1 as there is a slight interaction between the two capacitors. If the dimensions given in Fig. 4-13 are used, there should be little or no difficulty in obtaining a virtual zero reading on the bridge indicator. If a deep null cannot be obtained, check the ground connection of capacitor C2 and the antenna center point to the boom. As a last resort, try a slightly larger capacitance value for C2 or change the position of the shorting bar on the gamma rod. The omega match generally is very easy to adjust and usually no trouble is encountered. The line swr should be checked at the transmitter end of the coaxial cable.

When working on a tower or pole, it is sometimes difficult to carry or handle test equipment because of the strained conditions. This is particularly true of test equipment powered from the 120-volt ac lines. When making impedance measurements or adjustments, it is *not* necessary to *physically* attach the instruments to the point being measured or adjusted. In Section 2-8 we stated that a transmission line equal to a half wavelength, or multiples of a half wavelength, always repeats its far-end termination at the input end. The R-X bridge and signal source may be connected to the antenna feedpoint through an *electrical* half-wavelength line, measured and cut as outlined in Section 3-4. The test line can be of *any impedance* but its length must be exactly one or more half-wavelengths long. The smaller RG-58/U or RG-59/U coaxial lines make good test cables. The signal is applied through the test cable and the matching adjustments are carried out as described above. To keep from climbing up and down the tower, have another person on the ground to make any required adjustments of the bridge or the signal level. A pair of binoculars will permit indicator readings without the necessity of climbing down the tower.

The adjustments on gamma and omega matching systems when used on vertical antennas and towers are carried out in the same manner as described for horizontal antennas. In shunt-fed arrangements, the gamma rod may actually be a length of No. 12 copper wire attached to the antenna element with a stainless-steel sheet metal screw at a point about one quarter up the antenna from the base and drawn tight to an insulator mounted about 3 or 4 inches away from the base. See Fig. 4-9C. The gamma and omega capacitors, whether used on horizontal or vertical arrays, should be placed in a plastic or metal box and sealed against the entry of moisture and insects. The omega match is used more often for vertical antenna systems.

4-5. UNBALANCED-TO-BALANCED MATCHING SYSTEMS— TUNED AND UNTUNED

In the early days of amateur radio, the use of spaced, open-wire transmission lines with relatively high impedance was common. Such lines are still used today, particularly where losses in long runs of coaxial cable could become excessive. The matching devices used to feed balanced lines are commonly known as *antenna couplers* or *antenna tuners*. Although most of the old-timers are familiar with the methods of loading and matching transmission lines with antenna couplers, it is easy to damage interconnecting coaxial cables and accessories such as low-pass filters unless the proper matching procedures are used. We wish to emphasize that the following discussion is an impedance-matching procedure, even though a tuned circuit is involved.

The antenna tuner shown in Fig. 4-14 was used in the author's station for many years. The output-coil taps are adjustable to match parallel-wire transmission lines from 300 to 600 ohms. This tuner arrangement is especially suited to the station that uses a parallel-wire fed antenna on one or two low-frequency bands only and uses coaxial-line fed antennas on all of the other bands. Inductance L1 consists of 28 turns of No. 12 enameled copper wire wound on a 3-inch bakelite form. The spacing between turns is equal to the wire diameter. The enamel insulation is very carefully scraped away at a point on every other turn and a lug is soldered to the bare wire forming a line of connection taps across the coil. The tuning capacitor is a dual 200–200-picofarad variable type. For low-power, single-sideband transmitters or transceivers in the 100-watt range, C1 may be an old broadcast-receiver tuning capacitor with the sections connected in parallel. The link-resonating capacitor, C2, is a broadcast-receiver type with sections connected in parallel to give a maximum capacitance value of about 500 pF. Link coil L2 consists of 5 turns of No. 12 copper wire closewound over the center of L1. The entire assembly is constructed on a 12- × 17- × 3-inch aluminum chassis.

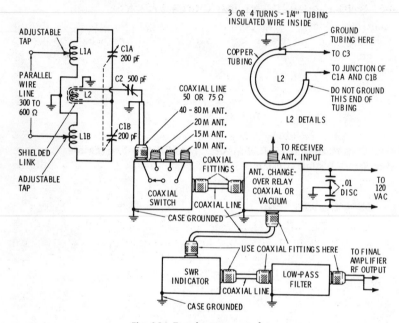

Fig. 4-14. Tuned antenna coupler.

To adjust the tuner, connect the output terminal of the R-X bridge to the tuner coaxial input cable. Apply the test signal to the bridge input and adjust the bridge resistance dial to 50 or 75 ohms, depending upon the impedance value of the tuner input cable. The bridge reactance dial is set at zero. Now, adjust tuner capacitor C1 for a null on the bridge indicator. The parallel-wire transmission line from the antenna is connected to two taps on coil L1 equidistant from the center. Alternately adjust capacitors C1 and C2 for the deepest possible null on the bridge indicator. Try moving the transmission-line connections out toward the ends of coil L1, one tap at a time, each time readjusting capacitors C1 and C2 for the deepest null indication. Eventually, a combination of L1 tap connections and C1 and C2 adjustments will be found that will produce a virtual zero reading on the bridge indicator. The dial and tap settings should be recorded for future reference.

This same circuit may be used as a phase-inverter to feed two antenna elements 180° out of phase. Coaxial lines of any impedance may be connected to the coil taps. The general adjustment procedure is the same as for a parallel line.

4-6. LENGTH-ADJUSTMENT MATCHING SYSTEMS

The length-adjustment method of matching the antenna feedpoint impedance to a coaxial transmission line is used extensively in vertical ground-plane antenna systems, especially at the higher frequencies. With this method, the radiator element is made slightly longer or shorter than the length required for resonance. At resonance, the base feedpoint impedance of a quarter-wavelength ground-plane antenna or a quarter-wavelength vertical radiator worked against a good ground system will be about 36 ohms. If the antenna element is deliberately made longer or shorter than the resonant length, reactance (X_L or X_C) will be present which will raise the base feedpoint impedance. This can be verified by measuring the base impedance with the R-X bridge. If the antenna is made short, an inductance must be added in series with the antenna and ground to cancel the inherent X_C of the antenna element. If the antenna is made long, the inherent reactance will be inductive and a capacitive reactance must be added between the antenna feedpoint and ground.

In most cases, it is more convenient to make the antenna element slightly shorter than the resonant length and add X_L to the system for matching purposes. The X_L may be in the form of a small, air-wound, self-supporting coil about 1 to 2 inches in diameter made from bare No. 12 copper wire or from ¼-inch copper tubing for higher power requirements. For 40-, 20-, 15-, and 10-meter verticals, the antenna length is made about *5-percent shorter than the calculated resonant length*. The inductor for a 40-meter radiator will be 1.5 inches in diameter and consist of about 15 turns of No. 12 *bare* copper wire with spacing about equal to the diameter of the wire. For the 20-, 15-, and 10-meter bands, the coil will be 1 inch in diameter and consist of 8 to 10 turns of No. 12 bare copper wire spaced about one wire diameter apart. All of these coils are larger than required; after matching, the excess coil is cut off and discarded. The coil should be mounted in such a manner that the coaxial line may be attached without sharp bends or kinks. All connections between the coaxial lines, the inductor, and the antenna element should be soldered, if possible. When the antenna element is aluminum, the electrical connection to it should be made with a stainless-steel sheet-metal screw.

To match the antenna to the coaxial line, connect the R-X bridge and signal source to the antenna and L1 as shown in Fig. 4-15A. Connect a 3- or 4-inch piece of copper braid (outer conductor of RG-8/U cable) to ground and terminate the other end of the lead in a small alligator clip. Adjust the bridge resistance dial for the same value as the coaxial line impedance and set the reactance dial to zero. Next, move the clip connection along the

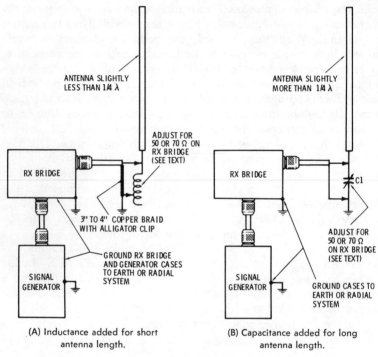

(A) Inductance added for short antenna length.

(B) Capacitance added for long antenna length.

Fig. 4-15. Connections for length-adjustment impedance matching method.

coil in the direction of the antenna until a null is obtained on the bridge indicator. Rotate the bridge reactance dial to check null depth. Continue to move the clip connection in each direction until the bridge indicator reads as close to zero as possible when the reactance dial is set at zero. The correct adjustment of the clip connection will be quite critical. A change of half turn or so may produce a great change in the null depth, especially on the 15- and 10-meter bands. Once the correct clip position is determined, remove the clip and solder the braid to the exact point on the coil where the clip was connected. Cut off and discard the excess coil turns.

If the antenna is made about 5 percent longer than the calculated resonant length, a variable capacitor may be connected in series with the antenna and ground, as shown in Fig. 4-15B. The R-X bridge is connected to the base of the antenna and adjusted as outlined above. Adjust the variable capacitor for greatest null depth on the bridge indicator. Remove the variable capacitor and, without disturbing its setting, measure its value on a capacitance bridge, or estimate its value as discussed previously. Now, select a 2500-volt mica capacitor with a value approximately equal to the measured or estimated value of the variable capacitor and connect it between the base of the antenna and ground. Once the approximate value of required capacitance is determined, several fixed capacitors with this rated value may be tried, one at a time, until the deepest null is indicated. The inductance-loading method is generally much simpler to adjust for an exact match. The *long element length* may be determined by the expression:

$$L_L = \frac{243.36}{f\,(\text{MHz})}$$

where,
L_L is the length of the long antenna element in feet,
f is the frequency in megahertz.

The *short element length* may be determined by the expression:

$$L_S = \frac{214.50}{f\,(\text{MHz})}$$

where,
L_S is the length of the short antenna element in feet,
f is the frequency in megahertz.

Fig. 4-16 shows an alternate arrangement for adjusting L1 or C1. A reflectometer or swr indicator is connected between the antenna and the transmission line. L1 or C1 is then adjusted for the lowest swr indication.

The X_L value added to the short antenna will be about 58 ohms; the X_C value added to the long antenna will be about 79 ohms. These X values will give a good match to either 50- or 75-ohm coaxial cables. If desired, the approximate inductance value re-

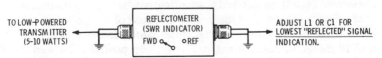

Fig. 4-16. Alternate arrangement for length-adjustment impedance matching.

quired to produce a 58-ohm X_L value may be calculated by the formula:

$$L = \frac{X_L}{2\pi f}$$

where,

L is the inductance required to load the antenna in henries,
X_L is the inductive reactance in ohms,
f is the frequency in hertz.

As an example, the approximate inductance value required to load a 7.2-MHz vertical radiator for a 50-ohm impedance match would be equal to $58/6.28 \times 7.2 \times 10^6$ henries or about 1.28 microhenries. Likewise, the C value required to load the long antenna element for a 50-ohm impedance match may be determined by the formula:

$$C = \frac{1}{2\pi f X_C}$$

where,

C is the capacitance required to load the antenna in farads,
X_C is the capacitive reactance in ohms,
f is the frequency in hertz.

4-7. THE STUB MATCHING SYSTEM

The stub matching method is very old and was used extensively in the early days of amateur radio when a "zepp" or a *pair of half waves in phase* were the latest thing in antennas. After the advent of coaxial transmission lines, the stub was largely relegated to the classroom and rarely used in practice. When the stub system is used with coaxial lines, there is some difficulty in *experimentally* locating the point on the line where the stub is to be attached. The length of both the "a" and "b" dimensions shown in Fig. 4-17 is critical. Using the "clipping method" to match a coaxial line to a load with an impedance higher or lower than the line impedance is very tedious to say the least. However, the stub length and its point of attachment on the coaxial line can be determined quite accurately by the method described subsequently.

The principal advantage of stub matching in phased and driven arrays is that it is not only an effective means of matching the load impedance to the line, but the electrical characteristics of the stub are stable and can be duplicated almost exactly in several units. As a result, it is much easier to adjust for and maintain proper phasing in an array which uses stub impedance match-

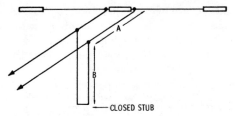

(A) Closed stub used with parallel transmission line.

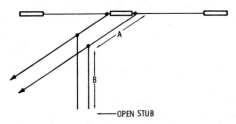

(B) Open stub used with parallel transmission line.

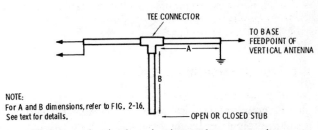

(C) Open or closed stub used with coaxial transmission line.

Fig. 4-17. Open and closed matching stubs.

ing as compared with systems using lumped-constant matching networks, such as the L-network.

The dimensions A and B of Fig. 4-17 are determined by the value of the line swr with no matching device between the line and its load. To measure the line swr, insert a *directional coupler,* or *monimatch* element, in series with the coaxial transmission line. Determine the line swr as accurately as possible. Now, refer to the curves in Fig. 2-17 (Chapter 2). Next, determine where the swr value intersects with the A and B curves. At the intersection, follow a vertical line to the bottom of the graph and read the length of the A or B section in decimal fractions of a wavelength. The physical length of the cable sections A and B can be calculated from the formulas previously given or from those at the back of this book. You will also notice that two separate sets of curves are shown—one for a shorted stub and the other for an open stub. The choice of an open or closed stub is determined by the value of the load impedance with respect to the line impedance value. If the antenna feedpoint impedance is *higher* than the line impedance, use the *closed-stub* configuration; if the antenna feedpoint impedance is *lower* than the line impedance, use the *open-stub* configuration. The antenna feedpoint impedance is determined by measuring it with the R-X bridge, as described previously. The A and B physical dimensions (in feet) are determined by the following expressions:

$$A = \frac{984VC}{fa}$$

where,

  A is the length of section A in feet,
  VC is the velocity constant for the coaxial cable used for the transmission line and the stub,
  f is the operating frequency in megahertz,
  a is the length of section A in wavelengths.

and:

$$B = \frac{984VC}{fb}$$

where,

  B is the length of section B in feet,
  b is the length of section B in wavelengths.

The velocity constant for standard "A" dielectric cables is 0.66, and for foam-dielectric cables it is 0.79.

Once their dimensions in feet are determined, the stub sections A and B are made up from the same type of cable as that of the transmission line. The electrical connections between the transmission line and the A and B stub sections are made by means of a coaxial tee connector. The bottom end of the stub, whether it is open or closed, should be sealed with plastic tape to prevent entry of moisture.

The input impedance at the tee connector may be measured with the R-X bridge. If the A and B sections are made an inch or two longer than the calculated length, the sections may be trimmed to an impedance value closely matching the line impedance by using the R-X bridge. In a phased array, all of the stub matching sections should have exactly the same dimensions.

The relative phase of the rf currents *at the elements* of a phased array may be adjusted by making *slight* changes in the length of the transmission line at the tee connector. Phase adjustments are covered in more detail in later chapters.

4-8. MISCELLANEOUS MATCHING SYSTEMS

In addition to the time-tested matching systems just described, there are several other impedance-matching arrangements including baluns, bazookas, and toroidal matching transformers. When the antenna feedpoint impedance is higher or lower than normal, such as when a folded element is used for a horizontal dipole, some of these devices are uniquely applicable. For some of these devices, there is no way to make impedance-matching adjustments once the device has been constructed. However, in the case of a device such as a toroidal matching transformer, the R-X bridge measurements will be useful in determining the proper turns ratios for matching specific impedance values.

It is often desirable to feed a *balanced* antenna element from an *unbalanced* coaxial transmission line. A good example would be a half-wavelength dipole antenna fed by a 70-ohm coaxial line. Even when the antenna feedpoint impedance equals the characteristic impedance of the transmission line, standing waves may appear on the outer conductor of the coaxial cable where it connects to one leg of the antenna. As a result, the coaxial line tends to act as a vertical radiator and, in effect, lengthens one leg of the antenna, creating a mismatch between the transmission line and the antenna feedpoint. Under these conditions, the entire system, antenna and line, may become frequency-sensitive. Some amateurs, baffled by this occurrence, keep trimming the antenna or transmission line until the system resonates within the amateur band.

The balun (a contraction of the expression *balanced to unbalanced*) is a device designed to eliminate this problem. When the balun is connected in series with the transmission line and the antenna feedpoint, *at the antenna,* standing waves are prevented from appearing on the outside conductor of the coaxial cable. Even though commercial baluns are frequently specified as having a 1:1 or 4:1 impedance transfer ratio, *the balun itself is not an impedance-matching device*. In other words, if there is an swr of 1.4 to 1 on the transmission line, and the condition is not due to standing waves on the outside of the line, the insertion of a balun with a 1:1 transfer ratio in the line will *not* improve the impedance match and the swr will still be 1.4 to 1. A balun with an impedance transfer ratio of 4:1 and fed from a 50-ohm line will have an output impedance of 200 ohms. The same balun fed from a 75-ohm line will have an output impedance of 300 ohms. From the basic balun configuration, however, there has been developed an impedance-matching transformer, generally known as an *rf transformer,* that is used for matching coaxial lines to various types of loads, especially the feedpoints of "short" vertical antennas with very low impedances.

The quarter-wavelength balun shown in Fig. 4-18 is constructed from coaxial line. It is perhaps the simplest type balun. A section of coaxial cable, a quarter wavelength long and of the same type as the transmission line, is placed parallel to the coaxial transmission line. The lower end of the quarter-wavelength section is grounded to the outer conductor of the coaxial transmission line. Together with the outer conductor of the transmission line, it forms a shorted, quarter-wavelength parallel-conductor balun with an impedance transfer ratio of 1:1. For amateur band use,

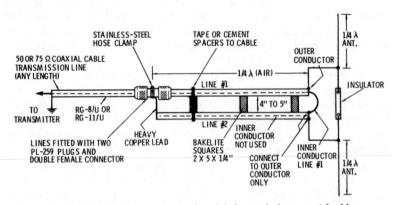

Fig. 4-18. Details of a quarter-wavelength balun made from coaxial cable.

the spacing between the parallel conductors should be such that the impedance is in the region of 300 to 400 ohms. When the system is operated at a frequency higher or lower than the resonant frequency of the antenna, the reactance of the shorted quarter-wavelength section will be opposite in sign to that of the antenna. Thus, if the system is designed for a frequency near the center of the band, the two reactances will tend to cancel each other when operating in the upper or lower portions of the band and a more constant impedance is presented to the coaxial transmission line. A system operated in this manner is said to be *broad banded.*

The rf transformer shown in Fig. 4-19 is a lumped-constant device whose electrical function is similar to that of the quarter-wavelength parallel-conductor balun. However, the broad-band rf transformer, unlike the quarter-wavelength parallel-conductor balun, is not confined to single-band operation. These broad-band transformers have very high bandwidth ratios (often as high as 20,000 to 1), and the same transformer may be used on all ama-

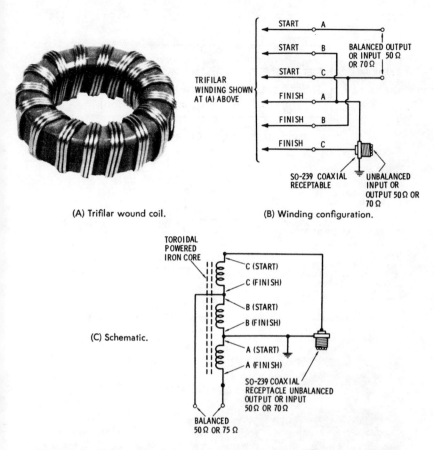

(A) Trifilar wound coil.

(B) Winding configuration.

(C) Schematic.

Fig. 4-19. Toroidal broad-band rf transformer with 1:1 balanced to unbalanced input or output.

teur bands from 160 meters well into the vhf region. Furthermore, the coils may be tapped to provide matching a 50- or 75-ohm coaxial line to antennas with impedances as low as 6 to 10 ohms and as high as several hundred ohms. The circuits illustrated in Figs. 4-19C and 4-20B are designed for feeding vertical antennas with various feedpoint impedances from standard 50- or 75-ohm coaxial line. The coils are either bifilar or trifilar wound (two or three simultaneous windings) and must be connected as indicated.

A commercially manufactured rf transformer is shown in Fig. 4-21. This unit, manufactured by Palomar Engineers, has impedance taps at 32, 28, 22, 18, 12, 8, and 5 ohms. The bandwidth is from 1 to 30 MHz for all taps except 12, 8, and 5 ohms. For the

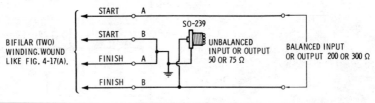

(A) Winding configuration.

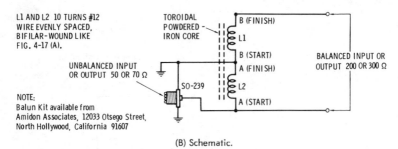

NOTE:
Balun Kit available from
Amidon Associates, 12033 Otsego Street,
North Hollywood, California 91607

(B) Schematic.

Fig. 4-20. Bifilar-wound balun transformer with 4:1 impedance transfer ratio.

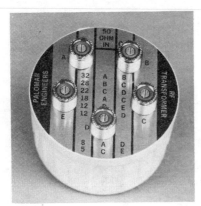

Courtesy Palomar Engineers

Fig. 4-21. Commercial rf transformer with taps at 32, 28, 22, 18, 12, 8, and 5 ohms.

three lower-impedance taps, the bandwidth is from 1 to 10 MHz. The rf power handling capabilities of the transformer are 2000 watts on cw and 5000 watts pep on single sideband. The core is rf ferrite material, the wire is Teflon insulated, and the waterproof PVC case is epoxy encapsulated. The coaxial connectors are standard uhf-type SO-239 chassis receptacles. The rf loss is less than 0-1 dB when operated into a matched load. The transformer is 3½ inches in diameter and 2½ inches high.

The following discussion will illustrate the use of the rf transformer. When used with a single-element, quarter-wavelength vertical antenna with a feedpoint resistance of about 36 ohms, a good match will be obtained by connecting the 50-ohm line from the transmitter to the 50-ohm input terminal and connecting the antenna to the 32-ohm tap. When using the transformer with short vertical antennas, the feedpoint resistance will be lower

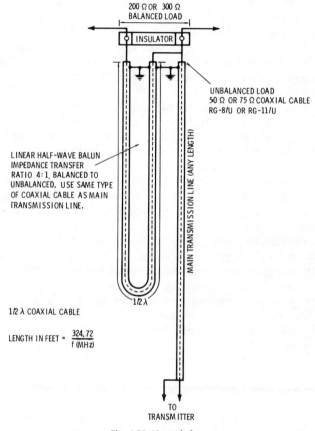

Fig. 4-22. Linear balun.

and the antenna must be resonated with a loading coil to tune out the inherent capacitive reactance. At resonance, a 0.2-wavelength antenna, for example, will have a feedpoint resistance of about 18 ohms. To match a 0.2-wavelength antenna to a 50-ohm transmission line, connect the transmission line to the 50-ohm input terminal and the antenna to the 18-ohm tap on the rf transformer. In practice, the feedpoint impedance is measured with the

R-X bridge to determine the actual value and the antenna is then connected to the tap which is closest to the measured value. The main advantage of the rf transformer is its simplicity and the fact that it has no critical adjustments.

The linear balun shown in Fig. 4-22 provides equal and opposite voltages, balanced to ground, from the inner conductors of the main transmission line and the half-wavelength phasing section. The impedance ratio is 4:1 stepped-down from the balanced to the unbalanced side. This type of balun is useful for matching a 75-ohm unbalanced transmitter output to a 300-ohm balanced line or for matching a 50- or 75-ohm coaxial unbalanced line to a 300-ohm balanced antenna such as a folded dipole or T-match arrangement. The greatest disadvantage of the linear balun is that it is a single-band device.

The balun transformers shown in Figs. 4-19 and 4-20 are wound on high-frequency, ferrite toroidal cores 2.4 inches in diameter with a permeability of 40. Such cores are advertised regularly in the amateur radio magazines. The configuration shown in Fig. 4-19 is a balanced-to-unbalanced transformer with an impedance transfer ratio of 1:1. It would be useful for feeding a half-wave horizontal dipole from a 75-ohm coaxial line. The configuration shown in Fig. 4-20 is a balanced-to-unbalanced transformer with an impedance ratio of 4:1 from the balanced side to the unbalanced side. Unlike the linear balun, the toroidal balun transformer is usable over a very wide frequency range, possibly over the entire hf amateur bands and well into the vhf range. The toroidal construction also has the advantage that the field of the transformer is almost wholly contained within the core and, therefore, is less likely to cause undesired coupling, radiation, and other problems associated with solenoid-type coils.

Again we wish to point out that, contrary to popular belief, the balun is not an impedance-matching device within the usual meaning of the term. The balun exhibits a definite impedance *transfer* characteristic which is usually fixed. However, it is necessary to adjust the line or antenna impedance, usually the latter, to a value near the impedance value that the transformer will match. Commercial rf transformers of all types will generally match a wide range of load impedances to a 50- or 75-ohm cable, but the transformer is not an easy correction for antenna problems caused by sloppy design. The general procedure in transformer matching is to select an appropriate output tap, depending upon the antenna length, etc. Then resonate the antenna by adjusting its length, or by adding X_L or X_C as previously discussed and adjust the antenna loading to the point where its feed-point impedance, measured with the R-X bridge, closely approximates one of the output impedance taps on the transformer. The line swr should always be checked before high-power operation is attempted. In phased-array systems, remember that all impedance-matching devices—linear transformers, baluns, etc.—will cause a phase shift, or delay, of the signal across the device. In multielement arrays, each element must be fed with its own matching device and the characteristics of all such matching devices must be identical in all respects.

SELF-EXAMINATION

Here is a chance to see how much you have learned about impedance measurements and matching procedures. These exercises are for self-testing only. Answer true or false.

1. The ultimate objective in impedance matching is to have the transmitter "look" into a pure-resistance load.

2. The simplest, and probably most often used, impedance-matching circuit is the L-network.

3. The L-network is used only to match high-impedance antennas to low-impedance transmission lines.

4. With the L-network random-wire antenna tuner, the controls are adjusted for lowest reading on the tuning-indicator meter when SW1 is in *reverse* position.

5. The pi network is useful for matching relatively low impedances to each other.

6. The gamma matching system can be used to match a transmission line to a vertical radiator or tower.

7. The main advantage of the omega or modified-gamma matching system is its ease in making gamma-rod length adjustments.

8. The purpose of the series capacitor in a gamma match system is to tune out the gamma stub inductance.

9. The balun is primarily an impedance-matching device.

10. The rf transformer will operate efficiently over the entire range of high-frequency amateur bands and well into the vhf region.

CHAPTER 5

System Design and
Measurement Procedures

In the preceding chapters our discussion has been confined to the characteristics of transmission lines, test-equipment design and construction, and impedance-matching problems and procedures. Now, we are ready to discuss the problems associated with the design of complete antenna arrays and ground systems, both simple and complex. In general, similar problems are encountered in all types of antenna systems. However, some of the undesirable system characteristics are more apparent in vertical systems, particularly switched and phased driven arrays. The three systems to be described here were constructed and tested in the order given. The author spent over 400 hours, almost an entire summer, experimenting with these antennas and their ground and radial systems. While it is not necessary to go to such great lengths with regard to ground-system evaluation in most cases, it is still very interesting and satisfying to *know* the characteristics of one of the most important components of the array.

5-1. GROUND SYSTEM EVALUATION WITH A REFERENCE ANTENNA

As mentioned previously, the quarter-wavelength monopole, when operated against an earth ground or radial system, is the basic reference antenna for evaluating the performance of the more sophisticated multielement arrays. It is unfortunate that the quarter-wavelength monopole is sometimes viewed by ama-

teurs as the least-desirable type of antenna. This lack of enthu-
siasm for the vertical monopole is largely due to a common mis-
understanding of its ground requirements and other characteristics. The vertical has been looked upon as an "instant" antenna
that is to be used in emergencies or in locations where there is
insufficient space to erect the supports for a horizontal antenna.
Because of this negative attitude, the installations are often sloppy
and the ground system may consist of nothing more than a rod
driven into the earth or a connection to a cold water pipe. While
the vertical antenna will perform to some extent under these difficult circumstances, the performance as compared with that of a
horizontal antenna system will depend largely upon the amount of
attention given to the installation of the radiator, the ground or
radial system, and the impedance-matching network. The radiator
should be installed in the clear, away from buildings and metallic
objects such as fences, water tanks, etc. The ground or radial
system used should be the best possible under the circumstances.
The method used by the author in evaluating the effectiveness of
the ground system used with several vertical antenna arrays will
now be discussed.

For many years, the author's greatest interest was DXing on
the 15- and 20-meter amateur bands. However, because of the
low sunspot activity and the resulting erratic behavior of these
bands, it was decided to construct some kind of directive antenna
system for the 40-meter band and try for DX contacts at these
lower frequencies. The first antenna tried was a half-wavelength
dipole erected about 40 feet above the earth. Various other types,
such as the inverted-V antenna, were tried with varying degrees
of success. Short- and medium-distance contacts were made with
ease but very little DX was worked. It soon became apparent
some kind of 40-meter beam antenna was desirable. The author
has used yagi-type horizontal beams on the high-frequency bands
for over 30 years and is thoroughly familiar with the details of
their electrical design, mechanical construction, and operation.
However, after a few inquiries with regard to the cost of a tower
and its installation, plus the cost of a heavy-duty rotator for a
40-meter beam, the idea of installing a phased and driven vertical
array became more and more attractive. It was finally decided to
make the design, construction, installation, and testing of this
array a detailed, long-term, amateur engineering project. The
work was carried out as described subsequently.

The logical first step in the design of the system was to select
the locations where the vertical radiator elements would be installed, taking into consideration clearance of trees, buildings,
and other obstructions and the directional characteristics of the

completed array. If interference from foreign broadcast stations operating in the amateur band is a problem, consideration should be given to placing the array so that a pattern null can be directed toward the interfering station. Systems which produce cardioid patterns are very effective in eliminating interference from strong stations when the pattern is switched so that the interfering signal is in the null position.

The second step was an evaluation study of the ground conductivity characteristics and the development of an effective radial system. Accordingly, the vertical monopole antenna shown in Fig. 5-1 was used in a study of ground- and radial-system effectiveness over a period of about six months. The radiator, constructed from 2-inch aluminum irrigation tubing and cut for resonance at 7.2 MHz, was installed in the clear and fed at the base with 50-ohm (RG-8/U) coaxial cable. At the beginning of

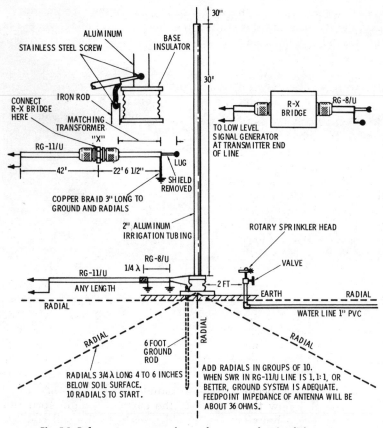

Fig. 5-1. Reference antenna used to evaluate ground and radial system.

the tests, only a 6-foot ground rod driven deep into the earth at the base of the antenna, was used as a ground system. Impedance measurements at the base of the antenna indicated an unexpectedly high resistance—in the order of 60 ohms. At this time California was in the midst of a severe drought and the earth was quite dry. Wetting down the earth around the base of the antenna made little or no measurable difference in the base resistance. At this point, it was decided to carry out a series of soil-conductivity tests. The topsoil at the author's home in the Sierra foothills of Placer County, California is decomposed granite and has a relatively high degree of conductivity, especially when it is damp or wet. The original exploratory tests were made with simply an ohmmeter and an ac voltmeter. Although the rf resistance of the soil will be different from the dc or 60-Hz ac resistance, the information obtained from the simple tests was very interesting and enlightening.

The antenna site was laid out in a crosshatch pattern with the crossover points spaced at intervals of 4 feet in each direction. The test probes were two ¼-inch diameter steel rods pushed into the soil to a depth of 30 inches. The test points were numbered and all resistance values were recorded for future reference. The value of recording test data as work on a system progresses cannot be overemphasized. In general, driving the rods deeper into the earth did not produce lower resistance readings. When the rods were laid horizontally in narrow trenches and covered with soil, the resistance between the rods was still not much different than when the rods were pushed into the soil.

In our case, the ohmmeter indicated very low resistance values between the two probes, even when the soil was dry. When the lawn sprinklers were left on overnight, the dc resistance between the two probes was less than one ohm. The resistance values obtained over the entire plot were remarkably uniform after the soaking. A similar set of resistance readings were taken by measuring the conductivity of the soil with 6-inch spike nails driven into the soil at the crossover points. When allowance was made for the amount of probe surface in contact with the soil, the readings taken with the spikes were close to those taken with the rods. From these observations it was concluded that, other than lightning protection, no advantages would be obtained through the use of a large number of ground rods or radials buried deep in the earth. In fact, it is the author's belief that less *dielectric* ground losses will occur when the radials are buried only about 3 or 4 inches below the surface of the earth. A single sprinkler turned on for an hour or so will add greatly to the soil conductivity.

After analysis of the test data, ten radials, each three-quarters wavelength long, buried 4 to 6 inches deep, and spaced 36° apart, were installed around the antenna. The far end of each radial was terminated in a 6-inch spike driven into the earth. At the antenna end, all of the radials were connected together and to the 6-foot ground rod. The base feedpoint resistance, after the sprinklers were left on overnight, measured about 50 ohms. We want to emphasize that the antenna was being used on the 40-meter band and signal strength reports were generally favorable. The reference antenna, at this time, was the half-wavelength horizontal dipole.

During the following several months, radials were added to the system in groups of ten and the base impedance of the antenna was measured and recorded after each change. The antenna resistance was lowered each time a group of radials was added until the total number of radials was 40. With 40 radials, the antenna resistance measured below 40 ohms with our homemade test equipment. No further reduction of the antenna feedpoint resistance was obtained when more than 40 radials were used. Interconnecting the radials at the far end and at intermediate points along their length produced no measurable change in the feedpoint resistance. The conclusion seemed to be obvious—for *our soil conditions*, we needed 40 radials in the ground system to stabilize the base resistance and produce normal operation of a quarter-wavelength, 40-meter vertical monopole antenna. Until the final impedance stabilization was obtained, the antenna had been fed from a 50-ohm line through an L-network. After stabilization, the feeder system consisted of a 75-ohm (RG-11/U) line from the transmitter to a 50-ohm (RG-8/U), quarter-wavelength linear matching transformer connected directly to the base of the antenna. The swr on the 75-ohm line at the transmitter was, for all practical purposes, almost unity and remained at that value during the several months that the antenna was in use. This antenna proved to be much more effective than the half-wavelength horizontal dipole reference antenna for medium and long distance contacts. The only ill effects noticed with the vertical antenna was a higher noise level, particularly summertime static (QRN). The field strength patterns were plotted and filed.

5-2. THE TWO-ELEMENT VERTICAL ANTENNA

As soon as the monopole was performing well, the author, like most hams, got the itch to experiment further with a more-complex vertical antenna system. After much study and on-the-air discussion with other amateurs, a second identical vertical monopole element was installed. The same number of radials were installed around the second element and measurements on both elements indicated identical feedpoint impedances. The original idea was to try to operate the two-element array on both 40 and 15 meters without making any changes in the feeder system. Accordingly, the spacing between the two vertical elements was made about 22 feet 6 inches—about 0.175-wavelength spacing on 40 meters and about 0.5-wavelength spacing on 15 meters. At this time, the 15-meter band was "dead" and no attempt was made to operate the system on that band. A switching network using two relays allowed the array to be operated with the cardioid pattern for east or west transmission and with an elliptical pattern north and south.

When relays K1 and K2 in Fig. 5-2 are not energized, the signal is fed to antenna A at 0° phase and it is fed to antenna B at −90° phase. This produces the east cardioid pattern shown in Fig. 5-3. When relay K1 is energized, the 0° signal is applied to antenna B and the −90° signal is fed to antenna A. A west cardioid is produced by this arrangement. The cardioid patterns are produced by *end-fire* operation of the two-element vertical array. With relay K2 energized as shown in Fig. 5-4, the signal is fed to

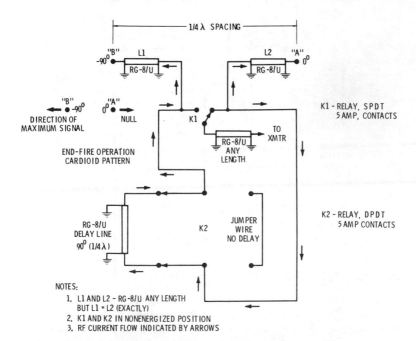

Fig. 5-2. Phasing arrangement for end-fire operations of two-element vertical array.

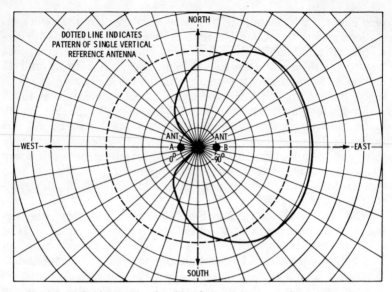

Fig. 5-3. Cardioid pattern produced by phasing arrangement shown in Fig. 5-2.

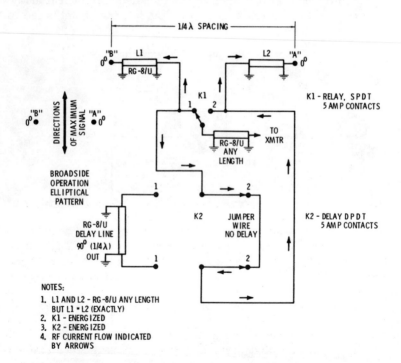

Fig. 5-4. Phasing arrangement for broadslide operation of the two-element vertical array.

both antennas at 0° phase. This arrangement is known as *broad-side* operation of the two-element vertical array and produces the elliptical pattern shown in Fig. 5-5.

The two-element vertical antenna has several desirable characteristics and a few that are undesirable. The maximum gain of the cardioid pattern appeared to be about 2.5 dB over the nondirec-

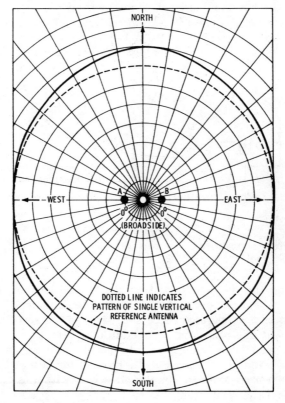

Fig. 5-5. Elliptical pattern produced by the phasing arrangement shown in Fig. 5-4.

tional pattern obtained with the single monopole. The front-to-back ratio on the cardioid pattern was very good; when a strong interfering signal was placed in the pattern null, it usually disappeared and was replaced by another signal from the opposite direction. The cardioid pattern is quite broad, about 135° at the half-power points, and most of the continental United States could be worked by pointing the cardioid pattern toward the east. Signal reports off the sides of the pattern in the direction of the half-power points indicated a signal strength greater than that obtained from the monopole antenna alone. The signal over a 60°

arc at the rear of the pattern was greatly attenuated as compared with the monopole.

When the two elements were driven in phase (0° phase difference), the line swr increased to the point where it was necessary to touch up the tank circuit of the linear rf amplifier each time that the pattern was changed. This annoyance was improved upon, but not entirely eliminated, by the insertion of an L-network at the junction of the main line from the transmitter and the phasing network, as shown in Fig. 5-6. Later, the element spacing was changed to 34 feet (approximately one-quarter-wavelength at 7.2

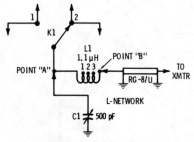

Fig. 5-6. L-network to reduce line swr in broadside position.

ADJUST C1 FOR MINIMUM SWR ON RG-8/U LINE WITH SWITCHING CIRCUIT IN BROADSIDE POSITION. TRY CONNECTING LINE TO VARIOUS TAPS ON L1. TRY CONNECTING C1 TO "A" OR "B" FOR LOWEST SWR. CUT OFF UNUSED PORTION OF L1.

MHz) and the line swr in the broadside position was improved further. It would appear that the two-element array could be close-spaced and operated successfully in small or restricted areas. However, it might require some additional attention with regard to the impedance matching in the broadside position. The line swr probably could be stabilized at the transmitter output by the use of a *transmatch* antenna coupler in series with the transmitter output and the main transmission line.

The second undesirable characteristic of the two-element vertical array is that when it is switched to broadside operation, the elliptical pattern has no front-to-back ratio. In the author's case, this was particularly annoying when trying to work stations in Alaska and Western Canada. Since the antenna sensitivity toward Southern California was equal to that toward the north, heavy interference (QRM) was frequently encountered. The transmitted signal, of course, was also very strong in the Los Angeles and San Diego areas.

In spite of these shortcomings, the two-element vertical array is a potent radiating system when properly installed and matched. Because of its relatively small size it can be used on the average city residential lot. When the elements are made from 2-inch aluminum as shown in Fig. 5-1, they are almost self-supporting. Three small-diameter nylon guy lines will keep the radiator upright, even in very strong winds. The appearance is also good, particularly when compared with the appearance of a 60-foot tower and a 40-meter yagi beam. In fact, one Southern California ham installed pulleys and nylon lines on his two-element array. On one element he displayed the American flag and on the other he displayed the California state flag.

5-3. A THREE-ELEMENT VERTICAL ANTENNA WITH TWO ELEMENTS DRIVEN

The two-element array described above was in use for a period of about three months. During that time, a number of DX con-

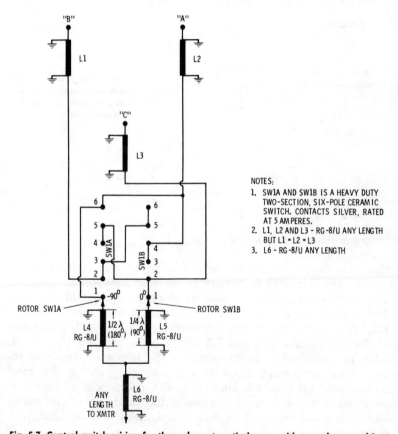

NOTES:
1. SW1A AND SW1B IS A HEAVY DUTY TWO-SECTION, SIX-POLE CERAMIC SWITCH. CONTACTS SILVER, RATED AT 5 AMPERES.
2. L1, L2 AND L3 - RG-8/U ANY LENGTH BUT L1 = L2 = L3
3. L6 - RG-8/U ANY LENGTH

Fig. 5-7. Control switch wiring for three-element vertical array with two elements driven.

tacts were made including Japan and South Africa. However, it was decided to add a third element as shown in Fig. 5-7 so that only the cardioid pattern would be transmitted in any direction. The pattern is switchable to six positions spaced 60° apart as shown in Fig. 5-8. The pattern is relatively broad, about 120° at the half-power points, and the forward gain appears to be over 3 dB with a rear attenuation of about 25 to 30 dB. The line swr after matching was about 1.2:1 in all positions. Since the cardioid pattern is unidirectional, interference from stations located in the direction at the rear of the pattern was greatly reduced. The layout for this antenna system is shown in Fig. 5-9. Fig. 5-8 shows

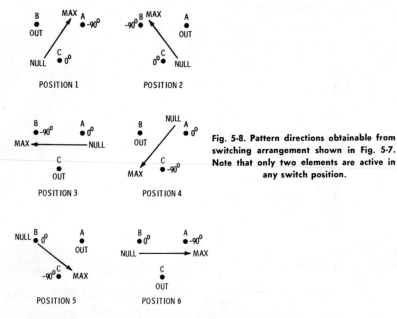

Fig. 5-8. Pattern directions obtainable from switching arrangement shown in Fig. 5-7. Note that only two elements are active in any switch position.

that only *two* elements are active in any given position of the switch. One element is always inactive in any switch position. The arrows indicate the direction of maximum signal strength. The system has a very broad impedance characteristic. With a 1.2:1 line swr at the resonant frequency of 7.2 MHz, the line swr at the band edges (7.0 and 7.3 MHz) measured less than 2.0:1. Also, no appreciable deterioration of the forward gain, front-to-back ratio, or pattern shape at *any* frequency within the band limits could be detected. Numerous DX contacts were made on cw in the lower-frequency portion of the band. This is an excellent antenna system but it has one drawback to be discussed subsequently.

As shown in Fig. 5-7, the system uses a two-section, six-pole, heavy-duty ceramic switch to select the direction of the pattern. Because of the complex switching arrangement, the circuit does not lend itself to remote control by relays. The switch was installed in a weatherproof housing mounted on an 18-inch high iron pipe placed on the lawn. In order to change direction of reception or transmission, it was necessary to walk out to the housing, open

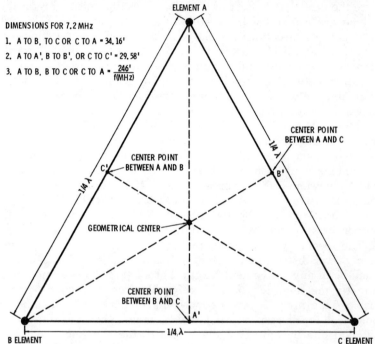

DIMENSIONS FOR 7.2 MHz

1. A TO B, TO C OR C TO A = 34.16'

2. A TO A', B TO B', OR C TO C' = 29.58'

3. A TO B, B TO C OR C TO A = $\frac{246'}{f(MHz)}$

ELEMENT A

CENTER POINT BETWEEN A AND C

CENTER POINT BETWEEN A AND B

GEOMETRICAL CENTER

CENTER POINT BETWEEN B AND C

1/4 λ

B ELEMENT

1/4 λ

C ELEMENT

Fig. 5-9. Layout for three-element vertical antenna system with two elements driven.

the door, and change the switch position by hand. Since the author is disabled, this proved to be a burden. This condition can be eliminated by installing the position switch and the phasing network inside the hamshack. When installed in this manner, the three coaxial lines from the switch and phasing network to the antenna elements must be of *exactly the same length*. Although more coaxial line is required with this arrangement, the total overall cost of the cables probably will still be less than the cost of good quality relays for a remote-switched system. With the switching network installed in the hamshack, there is also less chance of system trouble due to moisture problems. When remote-control relays are used, they must be protected against moisture and insects.

5-4. A THREE-ELEMENT VERTICAL ANTENNA WITH ALL ELEMENTS DRIVEN

The use of a power divider network and individual phasing and feed cables for each element of an array simplifies the remote switching problems. In the arrangement shown in Fig. 5-10, all three elements are driven simultaneously. The phasing, and therefore the directional effect of the system, can be selected by relays which are controlled by a switch panel at the operator's desk. The pattern directions for the various switch combinations are shown in Fig. 5-11. The system is more complex than those described previously and requires close attention to the power-dividing networks, the phasing networks, and the impedance-matching networks if optimum performance is to be obtained. Several arrays of this type have been described in the radio-amateur literature over the past several years. The prospective antenna designer is urged to explore the past issues of radio-amateur magazines for ideas before starting the construction of a complex, multielement, phased array. Fig. 5-12 shows the layout dimensions of the three-element, 40-meter, phased vertical array in current use at the author's station (W6TYH). The pattern directions use the formulas in Fig. 5-12 to determine the dimensions for the 15-, 20- and 75-meter bands.

The feeder system shown in Fig. 5-10 illustrates the phasing and power-distribution methods. Coaxial cables L1, L2 and L3 may be of any convenient length, but *all three cables must be of exactly the same length* and preferably from the *same batch* of cable. The three L-network matching sections, if used, must be identical in all respects. If possible, match the elements to the 50-ohm line by adjusting the length of the element rather than using the networks. When the L-networks are used, the phasing at the antenna element will be affected by the network. The possibility of obtaining proper phasing of the three elements is very remote when the networks are used unless measurements and adjustments are made as discussed subsequently. In the system shown, the swr in the 50-ohm line without the network was about 1.7:1. With the network in the circuit, and properly adjusted, the swr in the 50-ohm line was less than 1.2:1. Unless you have means of making phase measurements in the system, leave out the networks. The mismatch is not excessive and practically no difference in system performance will be observed.

Although the coaxial cables shown in Fig. 5-10 are designated RG-8/U and RG-11/U types, the smaller, less-expensive, RG-58 and RG-59 *foam-dielectric* cables may be used provided the swr is low—1.4:1 or better. The attenuation of the smaller foam-di-

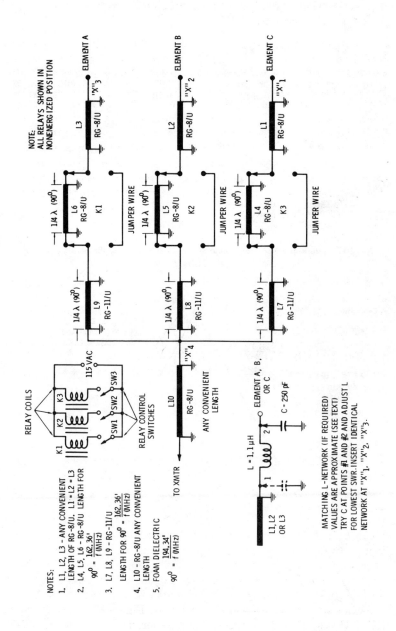

Fig. 5-10. Power-divider and phasing network for three-element vertical array with all elements driven.

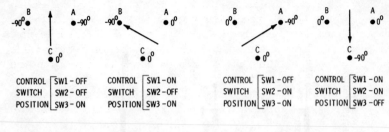

CONTROL	SW1 – OFF		CONTROL	SW1 – ON		CONTROL	SW1 – OFF		CONTROL	SW1 – ON
SWITCH	SW2 – OFF		SWITCH	SW2 – OFF		SWITCH	SW2 – ON		SWITCH	SW2 – ON
POSITION	SW3 – ON		POSITION	SW3 – ON		POSITION	SW3 – ON		POSITION	SW3 – OFF

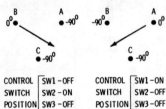

CONTROL	SW1 – OFF		CONTROL	SW1 – ON
SWITCH	SW2 – ON		SWITCH	SW2 – OFF
POSITION	SW3 – OFF		POSITION	SW3 – OFF

Fig. 5-11. Pattern directions for various switch combinations of the phasing network shown in Fig. 5-10.

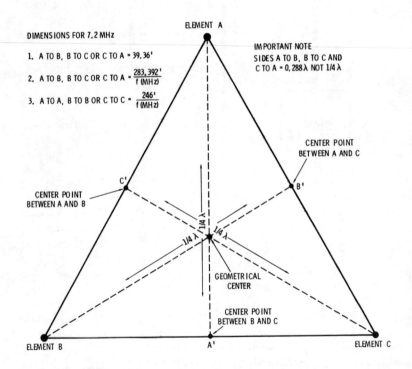

DIMENSIONS FOR 7.2 MHz

1. A TO B, B TO C OR C TO A = 39.36'

2. A TO B, B TO C OR C TO A = $\frac{283,392'}{f \text{ (MHz)}}$

3. A TO A, B TO B OR C TO C = $\frac{246'}{f \text{ (MHz)}}$

IMPORTANT NOTE
SIDES A TO B, B TO C AND
C TO A = 0.288 λ NOT 1/4 λ

ELEMENT A

CENTER POINT
BETWEEN A AND C

C'

CENTER POINT
BETWEEN A AND B

B'

1/4 λ   1/4 λ

1/4 λ   1/4 λ

GEOMETRICAL
CENTER

CENTER POINT
BETWEEN B AND C

ELEMENT B   A'   ELEMENT C

Fig. 5-12. Layout dimensions for the three-element array in Fig. 5-10.

electric cables at 7.0 MHz is about 0.8 dB per hundred feet when perfectly matched. If the antenna elements are matched to the 50-ohm lines, the measured swr in the system should be below 1.2:1 at the center of the band. If the swr is in this region, cable losses will be negligible. The main transmission line from the transmitter to the input of the power-divider section should be RG-8/U in any case. The velocity constant of the foam-dielectric cables is 0.79 as compared with 0.66 for the standard A-dielectric cable discussed previously. The proper formulas for both the standard A-dielectric and foam-dielectric cables are included in Fig. 5-10.

Many writers specify coaxial relays for switching the phasing sections in or out. While coaxial relays are desirable they are also expensive. The author has used both the expensive coaxial relays and the Potter & Brumfield KT11A miniature open-type relays. If the KT11A relays are properly installed, there is little or no significant difference in performance compared to coaxial relays on the 20-, 40-, and 75-meter bands. The KT11A is a double-pole, double-throw relay with 5-ampere gold-flashed silver contacts. It is rated for continuous duty at 120 volts ac. The armature insulation is glass silicone. Originally designed to switch rf powers up to 500 watts in a 300-ohm line, these little relays are ideal for switching the phasing components in or out of the antenna circuit. The leads to the contact points and the armature terminals are very short with low inductance. When the relays are mounted in individual aluminum shield boxes with coaxial fittings, the phase shift across the relay is negligible.

Probably the most difficult problems encountered in the design and adjustment of phased arrays are those of obtaining equal rf power distribution and the proper relative phasing of the antenna currents. The pattern shape, forward gain, and the front-to-back ratio depend on proper control of these factors. Power distribution and phasing in broadcast antenna systems are controlled by relatively complex networks which, while not beyond the technical capabilities of many amateurs, are generally not practicable because of a lack of measuring instruments and test equipment. Even in the amateur antenna systems phased with coaxial delay lines, lack of adequate test equipment is a serious problem. Often a relatively simple array, properly adjusted, will outperform a complex multielement array that has not been adjusted for optimum performance. In fact, unless the beginning amateur has access to proper test equipment, and possibly assistance from an expert, he should not attempt the construction and adjustment of the more elaborate arrays until he has gained some experience with the simpler systems. The ultimate three- or four-element system should be carefully planned to prevent waste of expensive coaxial cable and other materials.

The main transmission line from the transmitter to the power-divider section, as shown in Fig. 5-11, is RG-8/U. Do *not* use the smaller RG-58 cables for this line when operating at power levels above a few hundred watts. All of the antenna currents flow through this common transmission line and, if the swr is 1.5:1 or higher, the smaller cables may break down. The three quarter-wavelength (90°) power-divider cables are RG-11/U. The smaller RG-59 cables may be used here. However, if the foam-dielectric cable is used, you must use the proper formula for determining the quarter wavelength (see Fig. 5-10). The three power-divider cables are connected in parallel at the input end. Connection to the RG-8/U main transmission line is made through two coaxial T-connectors. The far end of each power-divider cable is fitted with a standard PL-259 plug. The relay shield boxes are fitted with SO-239 input and output coaxial chassis connectors. All fittings must make good electrical contact and be watertight. The cable fittings should be grounded. The easiest way to ground the fitting is to place a stainless-steel hose clamp around the fitting, insert the ground wire under the clamp, and tighten the clamp. Then wrap the entire connection with vinyl tape including about two or three inches on each side of the connection.

The quarter-wavelength (90°) phasing sections may be made from RG-8/U or the smaller RG-58 foam-dielectric cable. Again, when calculating the quarter wavelength for foam-dielectric cable, do not forget to use the correct formula. The phasing cables are fitted with PL-259 plugs at both ends. When measuring the quarter wavelength of cable, include the length of the plugs.

The 50-ohm cables from the relay boxes may be made from RG-8/U coaxial cable or RG-58 foam-dielectric coaxial cable. Both ends of each cable should be fitted with PL-259 plugs so that test instruments may be inserted in series with the line at the antenna terminal. All three cables must be of *precisely the same length* even though one or two lines are physically too long. The L-networks at the base of the antenna should be placed in waterproof aluminum boxes fitted with coaxial cable connectors. Tape all external connections as outlined in the foregoing discussion.

5-5. PRACTICAL IMPEDANCE AND PHASE MEASUREMENT PROCEDURES

The use of the R-X bridge and other test instruments has been discussed previously. The following discussion covers the application of these instruments to measurements in the antenna systems just described. When making these measurements, particularly phase measurements, always make certain that you have good electrical connections and that the instruments are stable. To check for electrical stability, place your hand on the oscilloscope or other instrument case and observe the indicator (scope trace or meter reading). If there is *any change* in the trace or meter reading, it *must* be eliminated before making adjustments in the antenna system. The usual correction is to connect all of the instrument cases together and to a good earth ground. Avoid the use of any kind of rf test cables; the transmission lines should be connected directly to the test equipment through coaxial fittings.

When measuring the base resistance (impedance) of an antenna element, it is not necessary to feed the test signal through the cable system from the transmitter. You already know that RG-8/U has a 50-ohm (nominal) surge impedance. Your problem is to make the antenna element appear as a 50-ohm "pure" resistance load when connected to the cable. Since the R-X bridge will indicate when the antenna resistance is 50 ohms, you need only to connect the output terminal of the bridge to the antenna and feed the test signal to the bridge input terminal as shown in Fig. 5-13. Before inserting any matching networks, measure the antenna base resistance and determine whether it also has a reactive component (X_L or X_C). The reader is referred back to Section 3-3 for detailed instructions. If the antenna feedpoint is reactive, as it most likely will be, you must decide whether or not to add X_L or X_C to the system. In cases where the base impedance measures just slightly higher or lower than the line impedance (say 40 or 60 ohms when using a 50-ohm line) it may be better to *not use* a matching network. The slight mismatch is of no great importance, while the insertion of the matching network will bring with it phasing problems. If the antenna base impedance is *much* higher or lower than the line impedance (say 30 or 80 ohms when using a 50-ohm line) then the correction network should be used. If a matching network is inserted at the feedpoint of one element in an array, an identical network must be inserted at the feedpoints of all other elements. Otherwise, the array cannot be properly phased. As pointed out previously, the antenna element should be resonated by adjusting its length whenever possible rather than doing it with added reactance. At any rate, after deciding whether or not to use a matching network, the actual swr at that point in the system should be checked by use of a reflectometer or swr meter as shown in Fig. 5-14. When checking line swr, the signal should be applied to the cable

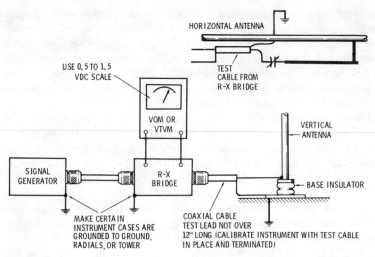

PROCEDURE:

1. ADJUST GENERATOR TO TEST FREQUENCY
2. ADJUST R AND X DIALS FOR <u>MINIMUM</u> INDICATION ON VTVM
3. IF READING IS NOT ZERO AT <u>MINIMUM</u>, ADJUST ANTENNA UNTIL LOWEST INDICATION IS OBTAINED WITH X DIAL AT 0 REACTANCE (CENTER ZERO)
4. READ IMPEDANCE FROM R DIAL

Fig. 5-13. Using the R-X bridge and signal generator to measure the impedance of an element in an array.

at the transmitter end. To check the swr, it may be necessary to use the transmitter since the signal generator rf output level is usually too low to give an indication on the swr meter.

When using RG-8/U and RG-11/U cables, the most serious effect of impedance mismatch anywhere in the system may be

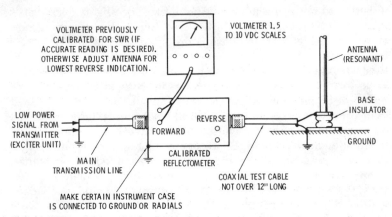

Fig. 5-14. Using a calibrated reflectometer for swr measurements and antenna adjustments.

detuning of the rf amplifier tank circuit or arcing of the relay contacts. With the smaller RG-58 or RG-59 cables, cable break-down may occur in addition to the other problems. Obviously, these conditions are more apparent when operating at the higher-power levels. The relays should *never* be "hot-switched" under any circumstances. To do so is to invite trouble from relay or cable breakdown, especially if the line swr is high.

The impedance and reactance (if any) may be measured at any point in the system where there are coaxial fittings. When insert-ing the R-X bridge or swr meter in series with a line other than at the antenna base, feed the rf signal into the main transmission line at the transmitter.

If the main transmission line swr changes when switching the pattern from one direction to another, the main line cable may be made perfectly *flat* (1:1 swr) by inserting an L-network at point "A," as shown in Fig. 5-6. Follow the same procedure as outlined in Section 3-3 for adjusting the L-network. Do not, un-der any circumstances, insert impedance-matching networks in series with the power divider or phasing cables. The L-network can be inserted at point "A" because the rf current has not yet been divided into its three components. After the current has been divided, insertion of any device will affect the phasing. Al-though we have discussed probable troubles when adjusting an-tenna systems, the measurements and adjustments should be sim-ple and relatively easy to do. All of the systems described are *broad-band* types and, if properly adjusted at the design fre-quency, should have a low swr even at the band extremes.

Phase measurements and adjustments on an antenna array are more difficult than impedance measurements and adjustments. Before making phase measurements on the array, the power-divider and phasing cables should be measured. The delay of the phasing cables is measured with the cable out of the system. However, the delay should be checked with the cables connected to the relay box. The relay coils are wired in such a manner that the phasing cable *is in the circuit* when the relay is *not* energized. Therefore, it is only necessary to measure the delay between the relay box input and output terminals. In this way we measure both the cable delay and the delay (if any) across the relay con-tact lead wiring. Delay measurements on the individual compo-nents will make the final system adjustments easier and will elim-inate some of the mystery involved in phasing driven arrays.

The use of the vectorscope or quadrascope discussed in Chap-ter 3 is recommended for the phase measurements. Each power-divider cable and each phasing cable (including the relay) should show a circular Lissajous pattern on the oscilloscope screen when the delay is exactly 90°. The cable delay should be measured before the coaxial fittings are soldered to the cable. If the cables are deliberately made two or three inches longer than calculated, they may be trimmed to exactly 90° delay. This is a tedious procedure, but when finished you will *know* that your cable delay is correct. Always use the same frequency for all measurements.

The next step is to make phase measurements at the antenna elements. Because of the effect of test leads and cables on the system, it is not possible to make direct connections to the antenna elements themselves. Fig. 3-22 shows two methods of taking rf energy from the element without disturbing its operation in the system. The large copper-tubing loop may be permanently installed at the base of the antenna for future phase measurements or the installation of a phase-monitoring system. The parallel pickup rods may be installed for temporary measurements. Both the loops and rods should be adjusted for equal pickup at all three elements of the array. Feed a *low-power* rf signal into the main line at the transmitter and measure the rf voltage at each pickup point with a low-range voltmeter (vtvm) and rf detector probe. Have someone observe the voltmeter while the rf power level is increased to obtain a reading. It is very easy to damage the meter if the power level is excessive.

After the pickup points have been adjusted for equal rf voltage, select one antenna element to be used for the phase reference (0°) angle. Apply the signal from this element to the horizontal plates of the vectorscope through a suitable length of coaxial cable. A 0° reference signal will be obtained from the element when its control relay is energized, thereby removing its delay cable from the circuit. Now, connect the lagging (−90°) element signal to the vertical plates of the vectorscope through a second piece of coaxial cable of the same type and *exactly* the same length. It will be necessary to use the transmitter, or exciter, as a signal source since the low-level signal generator will not develop the approximately 100 volts of rf necessary to produce usable patterns on the vectorscope. If the two antenna elements are actually running at a 90° phase difference, the vectorscope will display a circular pattern. If the measurements procedure has been carefully followed up to this point, and no matching networks were used at the antenna feedpoints, the phase difference between the two elements under test may be very close to 90°. If matching networks have been used, it is probable that some phase correction will be required. Before making any adjustments, however, check the delay between the 0° reference element and the second lagging element in the array. If both lagging elements show a similar amount of delay, and it is close to −90°, the array will probably function normally when the three relays are in this position. The direction of maximum signal transmission will be away from the 0° reference element and in the direction toward the lagging elements.

The next step is to select one of the elements that was previously lagging, use it as a 0° reference element, and measure the delay between it and the other two elements. The correct procedure is the same as that outlined in the foregoing paragraph. Continue this procedure, using each element in turn as the 0° reference element, until all pattern directions have been checked. It may be found that the phase shift in certain directions will be different from the others. As mentioned previously, the author has experienced great difficulty from a water-pressure tank located along one side of his array. It very definitely causes distortion of the pattern in three directions. However, since the tank cannot be removed, a slight adjustment was made in the cable system in an attempt to compensate for its effect. In spite of its known effect on the array, signal reports from the three affected directions are about the same as those from the other three directions. Several months of experience with this array (and others) has proved that, even though it may not be perfectly phased, it is still a potent radiator because of its low vertical-radiation angle.

If the phase difference between the antenna elements is much greater than 90°, it is probably due to the use of the L-networks for impedance matching. It is advisable to make the coils for the matching network from 1/8- or 1/4-inch copper tubing so that they will be self-supporting. The coils should be at least 2½ inches in diameter with the turns spaced about ½ inch apart. For the 40-meter band, about 10 turns of copper tubing 2½ inches in diameter and spaced about ½ inch between turns will be about right to start. Adjust the tap on the coil for the proper impedance match, as outlined in Section 4-1 and then cut off the unused portion of the coil, leaving about ¾ turn of unused coil in front of the tap position.

All three L-network coils should be identical in every respect —diameter, number of turns, and spacing between turns. After matching the antenna impedance to the transmission line, the three taps should be at the same position on each coil. If one element tends to match at a different point on its coil, the element length is slightly different from the others or something in its field is affecting its resonant frequency. One element in the author's array is near an aluminum roof that caused this problem. If possible, adjust the length of the affected element to restore resonance.

Now, connect the vectorscope to the pickup coils or rods as outlined previously and adjust the tap positions on the network of the two lagging elements until circular Lissajous patterns are obtained on the vectorscope screen. Use each element, in turn, as the 0° phase reference and then adjust the tap on the other two coils for the best circular display.

If matching networks are not used, or if difficulty is encountered in obtaining correct phasing in one direction, the coaxial lines from the relay boxes to the antenna feedpoints may be adjusted in length to produce corresponding changes in phase. Do not change the length of the original lines. These lines have been fitted with PL-259 plugs at the antenna ends so that a small length of coaxial line may be *added*. If the same amount of cable is added to each line, the phase difference is not changed. The phase shift, or delay, in an RG-8/U cable at 7.2 MHz is about 4° per foot. If we add a one-foot extension to *one* line, the current in the element to which it is attached will lag the reference element by about an *additional* 4°. By adding a two-foot extension to *all three* lines, we can trim the extensions and bring the system into the proper phase relationship. Measurements should be made with the control switches in different directional positions. It is doubtful that perfect delay will be obtained in all positions but the results should be close to the ideal. In the foregoing procedure, it is assumed that the power-divider and phase-shift cables have been adjusted for correct delay. Changes in the lengths of these cables to correct system delay is not recommended.

The effectiveness of the phasing adjustments can be checked by selecting a strong signal at the front of the pattern (on the 40-meter band, a foreign broadcast station is ideal), get a reading from the receiver's S meter, and then switch the pattern in the opposite direction and again observe the S meter reading. For the author's array, the difference in the S meter readings between front and back is about 15 dB. If the array is not in the clear away from metal roofs, water tanks, power lines, and even buildings, the front-to-back ratio will not be the same in all directions. The forward gain as compared to a quarter-wavelength reference antenna is at least 6 dB. The gain at the pattern half-power points compared to the reference antenna is about 3 dB. The pattern at the half-power points is about 120° wide. From the half-power points toward the rear, the attenuation of the radiation field is quite rapid. When the system is properly adjusted, signals received from a direction at the back of the pattern will be very weak.

The field-strength measurements made before and after adjustment of the system should be logically planned and carried out in an orderly manner. Field-strength measurements made in the induction field of an array are usually meaningless and may be misleading. Keep several wavelengths away when making these measurements and make all follow-up measurements from the same spot. If the field-strength meter is insensitive, use a pickup antenna and feed the signal to the meter through an 8- or 10-foot length of RG-58/U or RG-59/U coaxial cable. The author uses an 8-foot CB antenna mounted on a wooden broom handle for portable purposes. For the recorded measurement data, he has several vertical-wire antennas hung from tree branches at the appropriate points. Better readings will be obtained when the vertical antenna is held high. Field-strength readings should be taken from at least four positions—one at the front-center of the pattern, one each in the direction of the half-power points, and one at the rear-center of the pattern. Readings should be taken at each position while the pattern is switched to each of its six directions. The plotted patterns resulting from these measurements are always interesting and often astonishing, to say the least. From a study of these patterns, it is possible to make improvements in the array. Always plot new patterns after any changes in the system. All patterns and accompanying data should be filed for future reference.

SELF-EXAMINATION

Here is a chance to see how much you have learned about system-design measurements, methods, and procedures. These exercises are for self-testing only. Answer true or false.

1. The evaluation of the ground-radial system and measurements made on a reference antenna will yield valuable data for the design of a directional vertical array.

2. If a cold-water-pipe ground is available, it is not necessary to install a radial system.

3. The number of radials used has very little or no effect on the base impedance of a vertical radiator.

4. Radials should be buried several feet below the earth's surface.

5. Vertical phased arrays have horizontally broad, low vertical-angle, radiation patterns especially suitable for DX but often do not show much directivity at medium and short distances.

6. The main problem with a two-element vertical array operated in both end-fire and broadside positions is that there is a change in the line swr when changing from one to the other.

7. The most practical vertical array using quarter-wavelength elements would be quarter-wavelength spacing with 90° or 135° phasing.

8. An L-network inserted in the main transmission line will reduce the transmitter tank circuit detuning when switching back and forth from end-fire to broadside operation.

9. A three-element vertical array, with two elements active and one element idle, will permit the transmission of a cardioid pattern in any of six directions.

10. In making impedance or phase measurements, check the ground connections by placing a hand on the instrument case. If the indicator reading does not change, the ground connection is good.

CHAPTER 6

Orientation Methods, Procedures, and Problems

The principal reason for the installation of a directional antenna system is to improve transmission and reception in the desired direction and to suppress or reduce it in other directions. Ordinarily, no great difficulty is encountered in the orientation of a horizontal beam with a rotator. In most cases, the horizontal array is simply rotated for best reception of the signal from the station with which contact is desired. Nearly always the direction of best reception is also the direction of best transmission. In the *steerable* array, however, the elements are physically fixed in position and the pattern is "rotated" by changing the relative phase of the rf currents flowing in the antenna conductors. These fixed-position arrays often are designed to have field patterns with deep nulls for suppression of signals from undesired directions. A cardioid is an example of such a pattern. Unless the array is oriented to place the unwanted signal such as that from a foreign broadcast station directly in the null, much of the effectiveness of the array will be lost. The orientation procedures given here are similar to those used by the military forces for installing their communications antennas.

Radio waves travel from the transmitting antenna to a distant receiving point by way of a great-circle route; therefore, it is important that you aim the front of the antenna pattern in the great-circle direction of the receiver. However, before you lay out the array orientation, two basic facts must be determined. First, a reference line must be established to determine the direction of *true north* from the point where your antenna is to be installed. Second, the great-circle direction of the remote point with which you wish to make contact or the direction of an unwanted signal which you wish to suppress must be established. Both directions may be accurately established without the use of scientific instruments or complex procedures. We shall now discuss these methods.

6-1. USE OF STARS FOR DETERMINATION OF TRUE NORTH

In the northern hemisphere, the *North Star* marks the position of the north geographic (not magnetic) pole with an accuracy of about one degree. This star is most easily located with the aid of the *Big Dipper* constellation. The Big Dipper consists of seven bright stars arranged in the shape of a dipper with a long curved handle as shown in Fig. 6-1. If an imaginary straight line is drawn

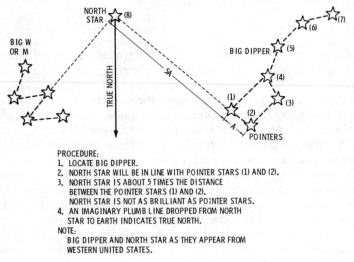

PROCEDURE:
1. LOCATE BIG DIPPER.
2. NORTH STAR WILL BE IN LINE WITH POINTER STARS (1) AND (2).
3. NORTH STAR IS ABOUT 5 TIMES THE DISTANCE
   BETWEEN THE POINTER STARS (1) AND (2).
   NORTH STAR IS NOT AS BRILLIANT AS POINTER STARS.
4. AN IMAGINARY PLUMB LINE DROPPED FROM NORTH
   STAR TO EARTH INDICATES TRUE NORTH.
NOTE:
   BIG DIPPER AND NORTH STAR AS THEY APPEAR FROM
   WESTERN UNITED STATES.

Fig. 6-1. Use of stars to determine true north.

through the two stars which form the side of the dipper opposite the handle and this line is projected about five times the distance between the two pointer stars, it will pass through the North Star.

To further aid you in locating the North Star, look for the *Big M* or *Big W* constellation which will be found to the right of the North Star. The center point of the "W" or the bottom point of the "M," as you prefer, points toward the North Star. If the night is hazy, it may be difficult to see the North Star because it is not as bright as the stars in the two reference constellations. Once true north is located, sight along the tops of two sticks or rods pushed into the ground to establish a *true-north reference line*. A nylon cord strung between the two sticks or rods will serve as a base to which all other directions can be referred. However, let us discuss another method of determining true north.

6-2. USE OF THE SUN AND A WATCH FOR DETERMINATION OF TRUE NORTH

This method is based on the fact that at exactly noon sun time, the bearing of the sun is due south in the northern hemisphere. Therefore, at exactly noon sun time, the shadow of a vertical mast or vertically suspended line will point toward true north. The only problem is that we must establish the instant of *true noon,* which is not necessarily 12:00 noon at your antenna site. To determine true noon, be certain that you have the correct time for your time zone (check with station WWV). Now, consult an atlas to determine the *longitude of your antenna site.* The time, as indicated on your watch, must be corrected one minute of time for each fifteen minutes of longitude (four minutes of time for each one degree of longitude). The amount of correction required will be determined by the difference in longitude between your antenna site and the meridian on which your own local time zone is based. In the USA, EST is based on the meridian at 75° of longitude; CST is based on 90° longitude; MST is based on 105° longitude; and PST is based on 120° longitude. Remember these are local *standard* times, not daylight-saving time. If your antenna site is located *east* of your time-zone meridian, the correction is *subtracted* from the local-time noon; if the location is *west* of the time-zone meridian, the correction is *added* to the local-time noon. As an example, the 120° meridian passes through a point near Reno, Nevada. When the sun is at "high" noon at this point, it is 12:00 noon anywhere in the PST zone but it is only 12:00 noon *sun time* on the 120° meridian. The author's station, W6TYH, is located on the 121° meridian. Therefore, true noon *sun time* at W6TYH will occur about *four minutes later* than in Reno. At a location one degree *east* of Reno, true noon will occur about *four minutes earlier* than in Reno.

To further confuse the reader, another correction must be applied for time of year because the earth varies in its orbit with respect to the sun. A graph showing these corrections for the twelve months is shown as Fig. 6-2. Again taking W6TYH as an example, the date is February 6th and we wish to establish a true-north reference line on that date. First, we add four minutes because we are one degree west of the 120° meridian. Then, we add another fourteen minutes for seasonal correction, or a total of *eighteen added minutes*. If our watch time (PST) is correct, true noon (sun time) will occur at W6TYH at 12:18 PM, February 6th of any year. As you can see, the calculation is very simple.

Now that you have determined the exact time of true solar noon, select a point on the antenna site from which to draw the true-north reference line. Suspend a vertical plumb line over this point. Now, at the exact time of solar noon (12:18 PM in the foregoing example) draw a line along the shadow of the vertical plumb line. Drive a stake into the earth at the end of the marked shadow line and another at the point where the plumb bob was positioned. Stretch a strong cord (nylon) between the two stakes to form a line that runs exactly true north and south. It is a good idea to project the line to the property limits and install markers for future reference. The reference line runs true north and south *all of the time,* not just at 12:18 PM on February 6th! Now, let us use the true-north reference line to enable us to point our signal accurately to *any* point on earth.

The next step is the orientation of the antenna array itself with respect to the true-north reference line. A world globe is probably on hand or can be obtained. To use the globe, you will also need an ordinary school protractor, two straight pins, and some lengths of white thread. If the globe has a center post at the north pole, tie one end of a length of thread around it and tie the other end of the thread to a straight pin stuck in the globe at the antenna site. Now, we have established a true-north reference line on the globe corresponding to our true-north reference line at the antenna site. Draw the thread tight between the two points on the globe so that it forms a straight line. Next, insert a straight pin in the globe at the point where the *undesired* station is located. Tie a piece of thread to the pin at the antenna site, draw it tight, and tie the other end of the thread around the pin at the undesired station site. We have now formed an angle with the two

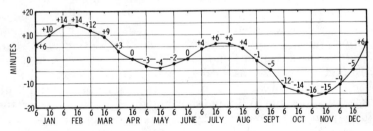

NOTE: ADD OR SUBTRACT MINUTES FROM LOCAL TIME (EST, CST, MST OR PST).
+ = ADDED   − = SUBTRACTED

Fig. 6-2. Standard-time correction chart for determining true noon.

pieces of thread which is the angle of antenna orientation with respect to a true-north reference line. Measure the angle with the transparent school protractor by placing the protractor so that the zero-degree line is over the thread that defines the true-north reference line. The zero point (or origin) of the protractor is centered over the pin that marks the antenna site. Most protractors have a small hole at the zero point. The protractor may be held in place by pushing the antenna-site pin through this hole into the globe. The angle formed by the two stretched threads can be read from the protractor.

The following discussion is based on the assumption that we want to attenuate the signal from the undesired station by placing it in the pattern null. Generally, if one undesired station can be definitely located in the null, other stations in the same direction will also be attenuated. Once the direction has been found, lay out the antenna *base line*. The base line is the actual line along which you desire minimum signal reception. The most accurate results in laying out the base line will be obtained with an engineer's transit. If you know a builder or someone that works for a construction company, he may lay out the base line for you in a few minutes. If a transit is not available, locate the spot where the *geometric center* of the array will be located (see Fig. 5-9) *on the true-north reference line.* At this point, push a rod or stake into the earth. Tie a nylon cord about 25 feet in length to the rod or stake. Place the protractor so that zero-degree line is over the cord that defines the true-north reference line and the zero point of the protractor is centered over the rod or stake at the geometric center of the array. Now, have someone walk the far end of the nylon cord back and forth while holding it taut until it indicates the same angle on the protractor that was obtained from the globe measurements. Once the correct angle between the true-north reference line and the base line has been determined, drive down a second stake or rod along the nylon cord and stretch the cord in a straight line between the two stakes. The nylon cord represents the base line of your antenna. We are now ready to locate the positions of the antenna elements.

The exact element positions will depend somewhat upon the type of array selected. The two-element array (Fig. 5-2) and the three-element array (Fig. 5-7) produce a switchable cardioid pattern along the line through the two elements. In the three-element array of Fig. 5-7 only *two* elements are active in any given direction with one element always out of the circuit. With these arrays, two of the elements, properly spaced, should be erected directly on the base line. The third element, if used, should be placed as shown in Fig. 5-9 to form an equilateral triangle. The null point and the "nose" (peak) of the reversible cardioid pattern will always be in the directions of a line through the two active elements. In the array shown in Fig. 5-12, however, all of the elements are active at all times. In this configuration, one of the lines A to A′, B to B′, or C to C′ should be chosen to coincide with the antenna base line. The radiated pattern from this array is not a cardioid, but signal pickup or radiation from the back of the pattern will be greatly attenuated.

Why do we select the null point in the pattern for orientation purposes? The reason is that the null point is much more sharply defined than the peak or "nose" of the pattern. Also, foreign broadcast and other foreign commercial service stations create severe interference, especially on the 40-meter bands. At the W6TYH location, a European broadcast station on 7.2 MHz is received with "twenty over nine" signals almost every day. All of the author's arrays have been oriented to place this station in the pattern null when transmitting toward Asia and the South Pacific. The fact that we chose 7.2 MHz as the center design frequency for our 40-meter arrays and that the European broadcast station is on the same frequency is not just a coincidence. We use the broadcast station to check the array performance, particularly the phasing and front-to-back ratio of the pattern. When the beam is switched for maximum signal from the broadcast station, an S-meter reading is noted. The beam is then switched in the opposite direction and the S-meter reading is again noted. If the array is properly phased, the difference between the two S-meter readings should be at least 12 to 15 dB. This method of checking the array performance is desirable because we are actually checking the antenna response on a constant-carrier DX station. Do not be surprised, however, if the maximum signal from the station is about the same level on two of the pattern "nose" positions. In the first place, the cardioid pattern is broad (over 100° at the half-power points) and second, the distant station may be received over more than one great-circle path at times.

The foregoing orientation procedures are equally applicable to all types of vertical and horizontal arrays. Accurate orientation is especially desirable for fixed unidirectional "V" and rhombic antennas. The accuracy with which you are able to predict the DX performance of the array will depend upon how much care you exercise in establishing the two reference lines.

6-3. RELATIVE AND ABSOLUTE FIELD-INTENSITY MEASUREMENTS

In our previous discussions of field-intensity measurements, it was assumed that the reader understood the use of the term *rela-*

tive. Simply, this means that all measurements are referred back to some reference level, usually the intensity of a signal radiated from a reference antenna. The use of relative field-strength measurements will enable the amateur to make changes or improvements in his antenna system relative to a reference antenna or to previous field-strength measurements. Relative field-strength measurements tell you how much better (or worse) your antenna performance is than it was previously or in comparison with another antenna. For commercial broadcast stations, where the field-intensity level at a given distance from the antenna may be specified in the station license, *absolute* field-strength measurements are almost always used. The serious amateur radio operator, interested in development or research, may wish to make absolute measurements of the radiated field intensity.

The field strength of a radio wave at any given distance in space from the transmitting antenna can be expressed in terms of the rf voltage induced in a wire of a specified diameter and length placed in that field. The rf voltage is usually measured in microvolts and, when a standard one-meter receiving antenna is used, the field intensity is expressed in basic units of *microvolts per meter.* The principal advantage of absolute field-intensity measurements is that the field-strength value may be determined independent of the type of transmitting antenna used.

It should be noted that it is not necessary to use an antenna exactly one meter (39.27 inches) in length. In actual practice, any length antenna can be used provided it is *calibrated* against a standard one-meter antenna. In addition to the calibrated antenna or one-meter antenna, you will also need a well-shielded communications receiver, a calibrated signal generator, an indicating meter (the receiver should have an S meter), and an attenuator calibrated in dB steps.

The following is a simplified field-strength measurement procedure suitable for amateur radio stations. The one-meter antenna is connected to the input terminal of the receiver which is tuned to the frequency of the transmitted signal. Observe and record the S-meter reading. Now, disconnect the one-meter antenna and apply the signal from the calibrated signal generator to the receiver input terminal. Adjust the signal level output from the signal generator until the receiver S-meter reading is the same as it was with the antenna signal. The absolute field strength of the transmitted signal in microvolts per meter will be the same as the level of the signal from the calibrated signal generator. If a calibrated signal generator is not available, your homemade low-level signal generator, previously described, may be calibrated against a standard laboratory-type signal generator. Most colleges that teach electronics and practically all radio and television stations have standard signal generators. A telephone call to the department head or chief engineer of the station is usually all that is necessary. He may even calibrate the generator for you or show you how to do it. It is not necessary to calibrate the generator over a wide range of levels as is done with a commercial laboratory-type instrument. If you have *one* level of output, perhaps maximum, that you can select with certainty every time, different *calibrated* signal levels may be obtained from the generator by the use of a calibrated attenuator.

The Barker & Williamson Model 371-1 wide-range attenuator shown in Fig. 6-3 is designed to provide step attenuation of low-

Courtesy Barker & Williamson, Inc.

Fig. 6-3. Barker & Williamson Model 371-1 wide-range attenuator.

level rf signals from signal generators, preamplifiers, or converters. Seven rocker switches provide attenuation of 1 dB to 61 dB in 1 dB steps. The switches are marked 1, 2, 3, 5, 10, 20, and 20 dB The sum of the actuated switches (IN position) gives the attenuation. With all switches in the OUT position, there is no measureable attenuation. The power capacity is 0.25 watt. The impedance is 50 ohms. The vswr is not more than 1.3:1 maximum from dc to 225 MHz.

If you desire to make your own attenuator, refer to any standard electronics engineering handbook, such as *Reference Data for Radio Engineers* published by Howard W. Sams & Co., Inc. To be most effective, the resistors and switch making up each attenuation step should be completely shielded from the others.

As mentioned previously, it is not necessary to use the standard one-meter antenna for measuring the absolute field intensity. When such an antenna is used, no correction factor will be necessary. However, when longer antennas are used for field-intensity measurements, a correction factor must be introduced because these antennas do not give field-intensity readings directly in microvolts per meter. To determine the correction factor, measure the field intensity of the radio signal using the standard one-meter antenna and record the reading. Now, measure the field intensity of the same signal using the longer antenna or *loop* and record the reading. Let us assume that we have obtained a reading of 100 microvolts with the standard one-meter antenna and a reading of 400 microvolts with the second antenna. The correction factor for the second antenna is equal to 400/100, or 4. Thus, if we pick up a signal on the second antenna which indicates a level of 600 microvolts, we must divide the 600 microvolts by 4 (the correction factor) to obtain the absolute field strength of 150 microvolts per meter. The correction factor for commercial field-intensity measuring equipment is supplied by the manufacturer. Commercial field-intensity measuring equipment designed for the range from 3.0 to 30.0 MHz generally uses a loop-type pickup antenna. In making ground-wave field-intensity measurements on amateur antennas, always place the pickup antenna in the same polarization plane as that of the transmitting antenna. For horizontally polarized antennas, the pickup antenna may be in the form of a dipole with the signal taken from the center and fed to the receiver through a short length of RG-59/U coaxial line.

6-4. COAXIAL-CABLE LINE-LOSS MEASUREMENTS

Most radio amateurs are familiar with the use of a dummy load for transmitter tune-up purposes. However, the dummy-load/wattmeter combination being produced by several manufacturers is not so widely used, mainly because many hams are not fully aware of the potentialities of this instrument.

In addition to its primary application, a dummy-load/wattmeter such as the Barker & Williamson Model 334A shown in Fig. 6-4 can be used to measure losses in coaxial cables and to check and adjust the calibration of vswr bridges. The frequency range of the Model 334A is from 2 to 300 MHz. The intermittent-duty power rating is 1000 watts and the nominal impedance is 52 ohms. The rf power input is by means of a hermetically sealed uhf connector on the rear panel.

This instrument is a completely passive device. The power-absorbing resistor is a structured monolithic 52-ohm noninductive resistor designed to present a constant-impedance load to the transmitter. The load resistor, which is tapped to provide an attenuated rf voltage to a diode, is mounted in a sealed steel container filled with a controlled-dielectric oil coolant. The schematic

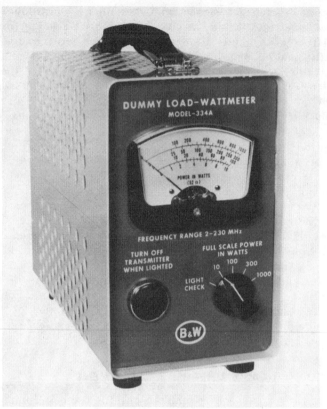

Fig. 6-4. Barker & Williamson Model 334A dummy-load/wattmeter.

diagram for the dummy-load/wattmeter is shown in Fig. 6-5. A temperature-actuated switch (S1) together with a panel-mounted light serves as an indicator of maximum safe operating temperature. The electrical connections to the sealed container are made through hermetic seals. The rf wattmeter is mounted outside the sealed container on the front panel of the instrument. A rectified dc voltage from the diode is fed to the metering circuit through suitable filters and a resistive attenuator to provide four full-scale power ranges of 0 to 10, 0 to 100, 0 to 300, and 0 to 1000 watts. Each of the four ranges has an adjustment potentiometer to calibrate its respective scale. Unless accurate power-measuring instruments are available, it is not recommended that these adjustments be attempted by the user.

Since the dummy-load/wattmeter is a power absorbing device, it will become hot in operation. Because of this heat, care should be exercised in the selection and use of the coaxial cable to connect the instrument into the circuit. For continuous use at maximum power, a high-temperature coaxial cable (RG-87A/U) is available. When using this cable, make no bends that are less than 5 inches in radius. Also be sure to connect the line cord so that the warning-light circuit will operate. The warning light may be tested by rotating the front-panel switch to the LIGHT CHECK position.

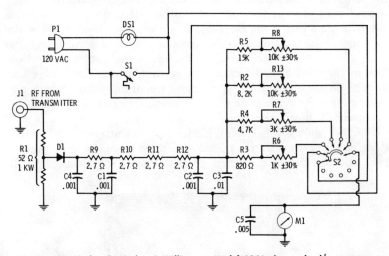

Fig. 6-5. Circuit for the Barker & Williamson Model 334A dummy-load/wattmeter.

The use of the dummy-load/wattmeter for measuring loss in coaxial cable is illustrated in Fig. 6-6. Connect the instrument to the transmitter, either directly or through a coaxial transfer switch. Turn on the transmitter and operate it on cw. Read and record the output power (W1) from the dummy-load/wattmeter. Next, insert the cable to be tested in series with the dummy-load/ wattmeter and the transmitter. Again operate the transmitter and

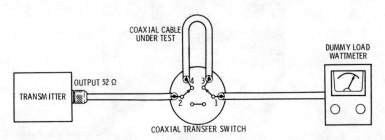

Fig. 6-6. Using the dummy-load/wattmeter to measure loss in coaxial cable.

record the power output (W2) from the dummy-load/wattmeter. The cable loss is:

$$\text{attenuation (dB)} = 10 \log \frac{W1}{W2}$$

The dummy-load/wattmeter can be used to check and adjust the calibration of swr meters described previously by connecting the dummy-load/wattmeter to the swr meter output in place of the antenna. Operate the transmitter. The bridge-indicating meter should read zero in the REVERSE position. If it does not, perform the balance adjustment as previously discussed. Next reverse the swr meter connections in the coaxial line and again read the bridge-indicating meter with the switch in the FORWARD position. The meter should read zero. If it does not, adjust the bridge balance to obtain a zero reading. Remove the dummy-load/wattmeter and connect the antenna to the swr meter. The swr meter will now accurately indicate the transmission-line swr.

Fig. 6-7 shows the Drake Model WV-4 wattmeter. This unit does not include a dummy load and must be used with the antenna or an external dummy load. The Model WV-4 has two full-

Fig. 6-7. Drake Model WV-4 wattmeter.

scale power ranges of 0 to 100 and 0 to 1000 watts. A similar
model, shown in Fig. 6-8, has two full-scale power ranges of 0 to
200 and 0 to 2000 watts. After the forward and reverse power readings have been recorded, the vswr may be determined from a
nomogram supplied with the unit. The schematic diagram for the
Model MV-4 wattmeter is shown in Fig. 6-9.

Fig. 6-8. Drake Model W-4 Wattmeter.

6-5. R-X NOISE-BRIDGE MEASUREMENTS

Noise-bridge measurements are a relatively recent development
among radio amateurs. In general, the noise-bridge principle is
based on the use of a wideband noise generator and an rf impedance bridge. An R-X noise bridge manufactured by Palomar Engineers is shown in Fig. 6-10. The two arms of the bridge are driven
equally by the noise generator through a broad-band ferrite-core
transformer. A third leg of the bridge has a calibrated variable
resistor (R) and a calibrated variable capacitor (C) in series.

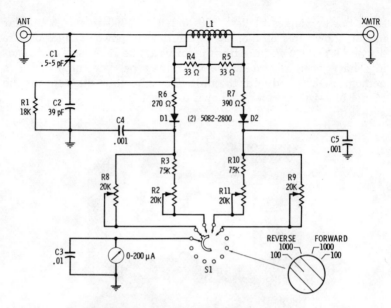

Fig. 6-9. Schematic diagram for Drake Model WV-4 wattmeter.

The antenna or other unknown circuit to be measured is connected as the fourth leg of the bridge. A communications receiver is used as the detector.

When R and C are adjusted for a null (minimum noise from the receiver), their dial settings can be read to find the resistance

Fig. 6-10. R-X noise bridge manufactured by Palomar Engineers.

and reactance of the unknown. As shown in Fig. 6-11, a capacitor is in series with the unknown so that capacitor C is at midscale when the unknown is a pure resistance. Thus both capacitive and inductive impedances can be measured. By tuning the receiver, the R and X of the unknown can be found at different frequencies. The useful range of the noise bridge is from 1 to 100 MHz.

To measure antenna resonance with the noise bridge, connect the antenna to the bridge UNKNOWN terminal and the receiver to the RCVR terminal. Connect a 9-volt transistor-type battery to the clips provided. Tune the receiver to the expected resonant fre-

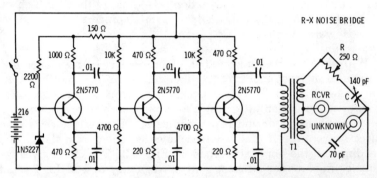

Fig. 6-11. Schematic diagram for Palomar R-X noise bridge.

quency and turn on the noise bridge. A loud noise will be heard. Adjust the R and X controls for a null (minimum noise). Note that the two controls interact and must be adjusted alternately until a deep null is obtained.

If the X control reading is on the X_L side of zero, the receiver is tuned to a frequency *above* resonance. If the X control reading is on the X_C side of zero, the receiver is tuned *below* resonance. Using the X control reading as a guide, retune the receiver and radjust the R and X dials for a null. With this procedure it is easy to find the resonant frequency of an antenna. At the resonant frequency (X = O) the R reading is the antenna resistance at the measurement point. If the measurement is made at a current loop (the center of a dipole for example), the indicated resistance is effectively the antenna radiation resistance.

Sometimes it is not possible to make the measurements at the antenna. Instead, the R-X noise bridge can be connected to the coaxial transmission line that feeds the antenna. This can be done by either of the following two ways:

1. If the transmission line is an electrical half-wavelength long, or some multiple of a half wavelength, then the readings

taken at the end of the transmission line will be exactly the same as though they were taken at the antenna. It must be remembered that there is only one frequency where the transmission line is exactly a half-wavelength long, so all measurements must be made at that frequency. Refer to Section 2-8 before making measurements with the half-wavelength line.

2. If the length of the transmission line is known, readings taken at the end of the line at *any* frequency can be converted using the Smith chart to find the antenna resistance and reactance. For information on the use of the Smith chart, refer to *Reference Data for Radio Engineers* published by Howard W. Sams & Co., Inc.

With the antenna connected as the unknown, its resistance and reactance at frequencies other than resonance can be found. At frequencies lower than resonance, an antenna appears as a capacitor and resistor in series; above resonance, the antenna appears as an inductor and resistor in series. The resistance is read directly from the R dial. The reactance is found from the X dial reading and the impedance chart in Fig. 6-12. The chart gives reactance in ohms for a measurement frequency of 1 MHz. To find the reactance at higher frequencies, divide the tabulated values by the frequency in megahertz.

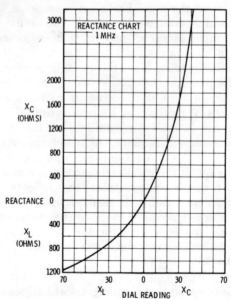

Fig. 6-12. Reactance chart used with Palomar R-X noise bridge.

To find the resonant frequency of a *series tuned circuit*, connect it to the UNKNOWN terminal of the R-X bridge. Set the R control to minimum resistance and set the X control to zero. Tune the receiver for a noise null. The X control can be used as described previously to determine whether resonance is above or below the frequency to which the receiver is tuned.

To find the resonant frequency of a *parallel tuned circuit*, connect a link coil with two or three turns to the UNKNOWN terminal. Bring the link coil close to the tuned circuit and use the procedure just described to find the resonant frequency. If the tuned circuit uses a toroid inductor, the link coil must thread through the toroid core.

The R-X noise bridge can also be used to determine the values of unknown capacitors and inductors. To do this, a standard capacitor (100-pF) and a standard inductor (5 microhenries) are used. To measure the inductance of a coil, connect it in series with the standard capacitor and find the resonant frequency. To measure a capacitor, connect it in series with the standard inductor and find the resonant frequency. The unknown inductance or capacitance may be determined by the appropriate formula:

$$L = \frac{25,330}{f^2 C}$$

and

$$C = \frac{25,330}{f^2 L}$$

where,
   f is the resonant frequency in MHz,
   L is the inductance in microhenries,
   C is the capacitance in picofarads.

With the resonant frequency known and either the standard capacitor or standard inductor in use, the value of an unknown inductor or unknown capacitor value can be calculated.

The noise bridge may also be used to determine the lengths of quarter-wavelength and half-wavelength transmission lines. To find the frequency at which a given line is an electrical quarter wavelength, connect it to the UNKNOWN terminal of the bridge and leave the other end of the line open. Set the R dial at zero and tune the receiver to the expected frequency. If the line is an exact electrical quarter wavelength, the null will be at X=0. If the receiver is tuned too low in frequency, the null will be on the X_C side of zero. If the receiver is tuned too high, the null will be on the X_L side of zero.

If it is desired to trim the line to resonance at a given frequency, the line should be disconnected from the R-X bridge. Short the UNKNOWN terminal with a jumper wire and adjust the X control for a null with the receiver at the desired frequency. Although the null will be at X=0, this method allows a precise setting to be made. Remove the short and reconnect the line but *do not* readjust the X control. Find the quarter-wavelength frequency by tuning the receiver to null. Trim the line slightly and retune the receiver. Repeat this procedure until the desired frequency is reached.

To find the resonant frequency for a half-wavelength line, the far end of the line should be short-circuited (instead of open-circuited as for the quarter-wavelength line). Then follow the same procedure as described for quarter-wavelength lines.

The R-X noise bridge has wide range controls (0 to 250 ohms and ±70 pF) which allow it to be used for many applications. Because of this, and because of manufacturing tolerances in the bridge components, the dials cannot be read as precisely as desired for some measurements. To find the precise setting for a given antenna resistance and to check the calibration of the bridge, resistors of known values can be connected to the UNKNOWN terminal. Carbon resistors of ¼- or ½-watt size are suitable for this purpose. They should be mounted in a PL-259 plug as shown in Fig. 6-13. The plug is then mated with the UNKNOWN receptacle of the noise bridge. Do not connect resistors to the UNKNOWN terminal with banana plugs or other open-wiring methods. Because of lead inductance with open-wiring methods, the X readings will not be correct.

Refer to the schematic in Fig. 6-11 and the impedance chart in Fig. 6-12. The R and C of the bridge circuit are the panel controls labeled R and X. When the X dial reads 0, C is 70 pF. A 70-pF fixed capacitor is in series with the UNKNOWN terminal. If the unknown series resistance and reactance are R_U and X_U, then at null:

$$R = R_U \text{ and } X_C = X_U + X \text{ (70 pF)}$$

The impedance chart shows this last relationship with reactance values given for $f = 1$ MHz. To find the reactance at higher fre-

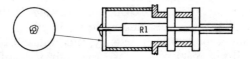

Fig. 6-13. Calibration resistor mounted in PL-259 plug.

quencies, divide the tabulated values by the frequency in mega-hertz.

The noise bridge is useful for measuring and adjusting trap-type dipole (all-band) antennas. Start with the highest-frequency band and measure the resistance and reactance as described previously. Adjust the center (or lower) section, if necessary, to resonate. Repeat this procedure for each successive lower band. The procedure is applicable to either horizontal or vertical trap elements.

To adjust yagi, quad, and similar types of beam antennas, connect the noise bridge to the driven element and tune the receiver to the operating frequency. Read the resistance and reactance. Adjust the antenna to resonance as required. With the noise bridge still connected to the driven element, nulls will be found at the resonant frequencies of the parasitic elements because of their close coupling to the driven element. The exact frequencies for these nulls will depend on the exact design of the beam. In one antenna, the resonant frequency of the director is 1.025 times that of the driven element, and the resonant frequency of the reflector is 0.98 times the driven element resonant frequency. The noise bridge is much more accurate for frequency measurement than the old standby grid-dip oscillator because the measured frequency is determined by the receiver which is usually quite accurately calibrated.

The author wishes to thank Mr. Jack Althouse, K6NY, Chief Engineer at Palomar Engineers, Escondido, California who kindly supplied much of the above information on the R-X noise bridge and its applications.

SELF-EXAMINATION

Here is a chance to see how much you have learned about orientation methods, procedures, and problems. These exercises are for self-testing only. Answer true or false.

1. To determine true north, use a good quality magnetic compass.

2. Radio waves travel from the transmitting antenna to a distant receiving point by way of a great-circle route. Therefore, a globe or special maps should be used as a guide to the layout of a directional array.

3. If we drop an imaginary vertical line from the North Star to earth, this line will indicate true north.

4. The North Star lies in an approximately straight line drawn through the Big Dipper pointer stars at a distance of about five times the spacing between the pointer stars.

5. When using the sun and a watch to determine true north, true noon will

depend upon the time of year and the location of the antenna site with respect to the standard-time meridian.

6. The antenna baseline is laid out with respect to the true-north reference line and a distant point on the earth's surface.

7. The null point of the pattern is used for orientation purposes because it gives greater accuracy.

8. The term *relative field strength* means that all measurements are referred back to a reference antenna, usually a dipole or monopole assumed to be operating under ideal conditions.

9. The term *absolute field strength* means that all measurements are referred back to a calibrated signal level, usually expressed in microvolts per meter.

10. Losses in coaxial cables are easily measured with a dummy-load/watt-meter.

CHAPTER 7

Useful Accessories

While the test equipment previously described in this book is perfectly suitable for most antenna measurements and adjustments, there are several additional accessories that will be found useful in both design and routine measurements of multielement antenna arrays. One of these accessories is a directional coupler designed especially for use with a vtvm, or a transistor voltmeter. The others include a pi-network line coupler for coupling a 75-ohm line to a 50-ohm line or vice versa, a half-wavelength filter which can be used to reduce harmonic radiation, or as a 180° phase-shifting (delay) network, a set of tuned autotransformers for converting the vertical antenna systems described in Chapter 5 to effective 14- and 21-MHz directional arrays, and a 5-watt rf signal source for the quadrascope phase measurements. The directional coupler will be described first.

7-1. DIRECTIONAL COUPLER FOR VTVM OR TVM

The directional coupler rf sampling element is an etched copper printed-circuit board as shown in Fig. 7-1. The unit is enclosed in a 5- × 3- × 2-inch aluminum box and is fitted with SO-239 uhf coaxial chassis connectors at each end. The printed-circuit-board element is designed to be an exact fit between the center conductors of the coaxial fittings, thus eliminating connecting leads. The two pickup strips are terminated at one end with a 68-ohm, 1-watt, carbon resistor and at the other end with a crystal diode. Two jacks are provided so that the forward or reverse resultant dc voltages from the diode rectifiers may be measured on the low dc-voltage ranges of the vtvm. If the coupler is carefully constructed, it may be inserted in series with a 50- or 75-ohm line *anywhere* in the antenna system to measure the line swr at that point. It is especially useful for determining the degree of matching that exists at the junction of the main transmission line and the power-divider or phasing sections. The coupler will permit swr measurements right at the antenna feedpoint or at any other point in the transmission line under actual operating conditions and is therefore a useful follow-up check on measurements made with one of the bridges previously described. In addition to the

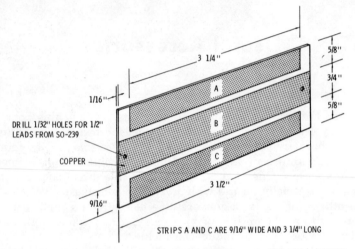

Courtesy M. L. Heineken, K5MNJ

Fig. 7-1. Printed-circuit-board sampling element for directional coupler.

inexpensive model designed for use with a vtvm, a complete directional coupler with a 0–100-μA dc indicating meter is shown in Fig. 7-2. The complete instrument is self-contained and does not require a power source of any kind other than the rf signal being measured. The instrument is calibrated for rf power measurements and is used by the author mostly for power-level adjustments in the 21- and 14-MHz directional arrays to be described later. The schematic diagram for the complete instrument is shown in Fig. 7-3.

The directional coupler is simple to construct and operate. The printed–circuit-board materials are available at most electronics parts stores or mail-order electronic parts suppliers. Using Fig. 7-1 as a guide, lay out the board as accurately as possible. Carefully cover all copper *to be left on the board* with printed-circuit resist tape. All exposed copper will now be etched away with printed-

(A) External view.

(B) Internal view.

Fig. 7-2. Directional coupler with built-in meter.

circuit etching solution. The author has had good results with the Dow Radio/Milo etching solution sold by most electronics parts stores. The board to be etched is immersed in the ferric chloride solution at a temperature of about 100° to 120° Fahrenheit. Use a plastic or glass tray and agitate the etchant back and forth over the board by rocking the tray gently. The etching process will

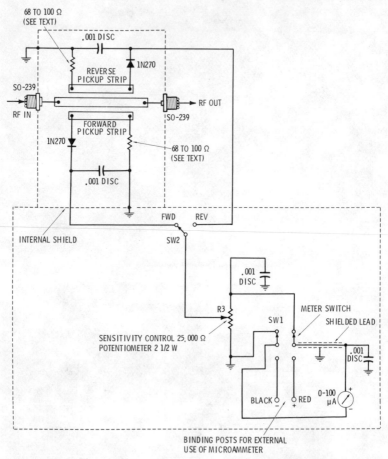

Fig. 7-3. Schematic diagram for the directional coupler shown in Fig. 7-2.

require about 45 minutes to one hour. After the undesired copper is etched away, wash the board thoroughly in running water and let it dry. Check for leakage or short circuits between the side and center copper strips with a high-range ohmmeter. *Caution:* The ferric chloride solution is dangerous to the skin and must be kept out of the reach of children. The author wishes to thank

Mr. M. L. Heineken, K5MNJ, who made up the boards used in the model units. The layout for the directional coupler is shown in Fig. 7-4.

To calibrate the directional coupler, adjust either the signal level or the meter sensitivity until the meter reads exactly full scale when switch SW2 is in the FORWARD position. Next throw the toggle switch to REVERSE position. If the transmission line is perfectly *flat* (has no standing waves) the meter should indicate

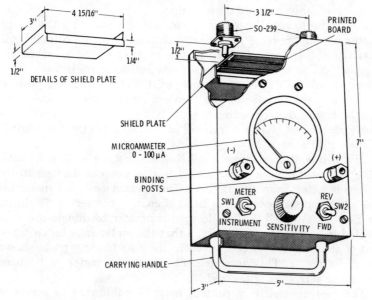

Fig. 7-4. Layout for complete directional coupler with built-in meter.

zero when placed in the REVERSE position. If it does not, adjust the value of the termination resistor on the reverse pickup strip until the lowest meter reading is obtained. The exact value of the terminating resistors will vary slightly with different pickup elements, but the meter usually will null at about 68 to 82 ohms. The best null was obtained by the author with the 68-ohm, 1-watt, *carbon* resistors. The resistors should be dressed away from the pickup strips and at right angles to reduce rf coupling between the resistors and the strips. Drill a small hole at the end of each pickup strip, insert the resistor lead through the hole, and then clip off the excess lead flush with the board. Solder the lead to the board using rosin-core solder and a low-wattage soldering iron. Drill two small holes for 6-32 machine screws in the back of the aluminum case, insert the screws in the holes, and hold them in place with a nut. Locate the screws so that the ground end of the resistor will practically rest on the tip of the machine screw. This arrangement will allow the leads from the resistor to the board conductor and the leads from the resistor to ground to be less than ⅛ inch long. Although the use of a calibrated dummy load for calibration of the directional coupler is recommended, the adjustments can be made with the antenna and transmission line if the swr on the line is low. If there are standing waves on the transmission line, the meter will not read zero but it will probably indicate a low value.

After you have adjusted the terminating resistor for the lowest meter reading, reverse the line connections to the directional coupler. Now, with switch SW2 in the REVERSE position, adjust the signal level or meter sensitivity for a full-scale indication. Next, place switch SW2 in the FORWARD position and adjust the position or value of the resistor on the forward pickup strip, if necessary, for the lowest meter reading in exactly the same manner as previously described. Keep a record of meter readings for various values of line swrs.

The self-contained directional coupler has the pickup or sampling element enclosed in a shield box to prevent damage to the microammeter from the rf field. The arrangement for mounting the sampling element on the SO-239 coaxial connectors is shown in Figs. 7-2B and 7-4. The dc microammeter terminals are connected to a dpdt toggle switch so that the meter may be switched out of the circuit and connected to the two binding posts shown near the meter. This arrangement allows the meter to be used for other purposes if desired.

The meter sensitivity control may be calibrated in terms of power output. To calibrate this control, connect a dummy load across the output terminal. Apply the rf power from the transmitter to the input terminal. The rf voltage across the dummy load may be read with a reasonable degree of accuracy using a vtvm and an rf detector probe. Begin with the vtvm on at least the 500-volt dc range and switch to lower ranges as necessary. Do not, under any circumstances, connect the meter without the probe across the dummy load. Also make sure that the diode in the probe will take the rf voltage across the dummy load without breaking down. The rectified rf voltage may be assumed, for practical purposes, to be equal to the rf rms voltage. This assumption will permit power-level measurements with an accuracy sufficient for most amateur-radio purposes. The power level may be calculated from the expression:

$$P = \frac{E^2}{R}$$

where,

P is the average power in watts,
E is the rectified voltage across the dummy-load resistor (as measured on the vtvm with a detector probe),
R is the resistance of the dummy load (usually 52 or 75 ohms).

The power dissipated in the dummy-load resistor may also be determined by measuring the rf current through the resistor with an rf ammeter. The meter, usually a thermocouple type, is connected *in series* with the line and dummy load. Power may be calculated from the expression:

$$P = I^2R$$

where,

P is the average power in watts,
I is the rf current flowing through the dummy-load resistor (as measured on the rf ammeter),
R is the resistance of the dummy load (usually 52 or 75 ohms).

7-2. PI-NETWORK LINE-TO-LINE COUPLER

Many times the amateur may wish to experiment with a certain type of antenna system but does not have coaxial line of the proper surge impedance. The pi-network line-to-line coupler shown in Fig. 7-5 will permit the use of 50- and 75-ohm coaxial cable in series at a much lower line swr than when the lines are connected directly together. Unlike a broad-band rf transformer, the pi-network coupler must be designed for a specific frequency and is a single-band device. However, it can be constructed quickly, usually from parts on hand. The component values for a given frequency may be determined from the expression:

$$C = \frac{159 \times 10^{-3}}{f \times X_C}$$

where,

f is the frequency in hertz or cycles per second,
X_C is the reactance in ohms equal to the line surge impedance,
C is the capacitance in farads.

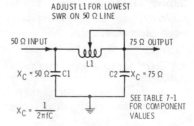

Fig. 7-5. Pi-network line-to-line coupler.

As an example, what capacitance value is required to present an X_C of 50 ohms at a frequency of 7.2 MHz? Remember that C will be in farads and must be multiplied by 10^{12} to obtain picofarads. The solution is as follows:

$$C = \frac{159 \times 10^{-3}}{7.2 \times 10^6 \times 50}$$

$$= 441.66 \times 10^{-12} \text{ farads}$$

$$= 441.66 \text{ picofarads}$$

A 440-picofarad capacitor with a high rf voltage rating will be suitable. For coupling between 50- and 75-ohm lines (such as RG-8/U and RG-11/U), calculate one capacitor value for 50-ohm X_C and the other for 75-ohm X_C. For a 7.2-MHz coupler, such as that shown in Fig. 7-5, the 50-ohm input capacitor is 440 picofarads; the 75-ohm input capacitor is 300 picofarads. The two capacitors used here are High Energy Corporation type HH58S ceramic transmitting capacitors with a peak voltage rating of 5000 volts.

Table 7-1. Component Values for Pi-Network Coupler in Fig. 7-5

Band	C1	C2	L1
3.5–4.0 MHz	820 pF	560 pF	16 turns No. 12 Copper wire, 1½" diameter, 8 turns per inch.
7.0–7.3 MHz	430 pF	300 pF	12 turns No. 12 Copper wire, 1½" diameter, 8 turns per inch.
14.0–14.35 MHz	220 pF	150 pF	8 turns No. 12 Copper wire, 1½" diameter, 8 turns per inch.
21.0–21.45 MHz	150 pF	100 pF	6 turns No. 12 Copper wire, 1½" diameter, 8 turns per inch.
28.0–29.0 MHz	110 pF	75 pF	5 turns No. 12 Copper wire, 1½" diameter, 8 turns per inch.

Note: The number of turns given for coil L1 is only approximate. Coil L1 must be adjusted for lowest swr on main transmission line. See text.

The component values for the pi-network coupler in Fig. 7-5 may be taken from Table 7-1. The coils are made with a few more turns than required and then adjusted by removing a turn or so at a time until the desired input and output impedances are obtained. The impedances may be measured with the R-X bridge or the swr may be adjusted with the directional coupler as previously described.

7-3. THE HALF-WAVE FILTER

The half-wave filter shown in Fig. 7-6 is somewhat similar in design to the pi-network coupler described above. In this case,

Fig. 7-6. A half-wave filter element.

however, the input and output capacitors must have an X_C equal to the line impedance. The units are designed for either 50- or 75-ohm lines and must not be used between lines of different impedances. Use the same capacitance formula given for the pi-network line coupler. The inductance values may be determined from the expression:

$$L = \frac{X_L}{2\pi f}$$
$$= \frac{X_L}{6.28 \times f}$$

where,
  f is the frequency in hertz or cycles per second,
  X_L is the reactance in ohms equal to the line surge impedance,
  L is the inductance in henries (multiply by 10^3 to obtain milli-
    henries and by 10^6 to obtain microhenries).

The half-wave filter components are enclosed in a 5- × 3- × 2-inch aluminum box fitted with standard SO-239 uhf coaxial connectors at the ends as shown in Fig. 7-7. In use, the units are connected in series with each other and with the transmission line. The author has used the half-wave filter as a 180° phasing section instead of using a half wavelength of coaxial line. When the filters were properly adjusted, no difference in performance could be detected between the filter and a half-wavelength coaxial line. Component values for half-wave filters for the high-frequency bands are given in Table 7-2.

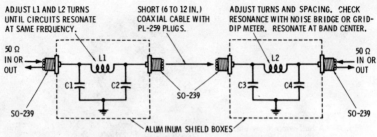

Fig. 7-7. Circuit for half-wave filter.

Table 7-2. Component Values for Half-Wave Filters in Fig. 7-7

Band	Capacitors C1, C2, C3, and C4 Capacitance	Coils L1 and L2			
		Inductance	Diameter	Turns	Spacing
80 Meter	850 pF	4.2 μH	1½"	13	8 Turns/Inch
40 Meter	440 pF	1.1 μH	1½"	7	8 Turns/Inch
20 Meter	220 pF	0.55 μH	1½"	5	8 Turns/Inch
15 Meter	150 pF	0.40 μH	1½"	4	8 Turns/Inch
20 Meter	110 pF	0.30 μH	½"	6	8 Turns/Inch

Notes: Capacitors C1, C2, C3, and C4 are 5-kV ceramic transmitting capacitors, High Energy
Corp. Type HH58S.
Coils L1 and L2 are air-wound using AWG No. 12 enameled wire.

7-4. TUNED RF AUTOTRANSFORMERS FOR 14- AND 21-MHz BANDS

The three-element vertical arrays described in Chapter 5 were designed primarily for 40-meter operation and were included primarily as subjects for the measurements discussions. However, they are effective long-distance (DX) radiators, not only on 40 meters but on the 14- and 21-MHz bands as well. When operated on the 14-MHz band, the 40-meter quarter-wavelength elements are a half wavelength long and the spacing is also a half wavelength. On the 21-MHz band, the length of the 40-meter quarter-wavelength element is three-quarter wavelength and the spacing is also three-quarter wavelengths. The feedpoint impedance in both cases is higher than either 50- or 75-ohm coaxial cable will match with a direct connection and still maintain a reasonably low swr. The tuned matching transformers described here will allow operation on the higher-frequency bands without any changes in the length of the phasing lines or the feeder system. The 14-MHz transformers will be discussed first.

The transformers shown in Fig. 7-8 are narrow-band devices adjusted for operation over the low-frequency end of the 14-MHz

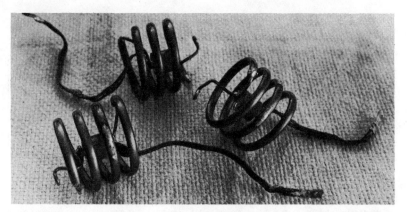

Fig. 7-8. Tuned rf autotransformer coils for operation of 40-meter phased vertical arrays on 14-MHz band.

band from 14.2 to 14.25 MHz. The schematic for the tuned rf auto-transformer is shown in Fig. 7-9. Each coil consists of 3½ turns of ¼-inch copper tubing, 2 inches in diameter, and spaced so that the coil is 3 inches long. Capacitor C1 is a 250-picofarad High Energy Corp. Type HH58S rated at 5000 volts peak. This capacitor will pass heavy rf circulating currents without heating or breakdown. A variable or fixed air-dielectric capacitor with the same

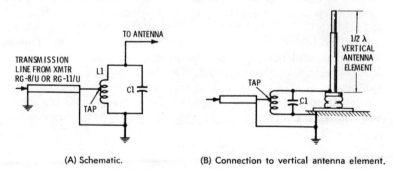

(A) Schematic.　　　　(B) Connection to vertical antenna element.

Fig. 7-9. Tuned rf autotransformer.

value may be used if desired. For the transformers shown, the resonant frequency is at 14.225 MHz and the system works well over the lower 50-kHz portion of the "phone" band. If complete coverage of the band is desired, C1 should be about 50 pF and the coil should consist of 4½ turns of ¼-inch copper tubing, 2 inches in diameter, and 3 inches long. With capacitor C1 fixed in value, the circuit is resonated by spreading or compressing the coil turns until resonance is indicated on a grid-dip meter or a wavemeter.

Table 7-3. Tuned RF Transformer Component Values for 15- and 20-MHz Bands

Band	C1	L1
15 Meters	25-pF, 7.5-kV Ceramic Transmitting Capacitor	3 Turns ¼" Copper Tubing 2¼" Inside Diameter, 2" Long Tap Approximately 1 Turn from Ground End
20 Meters	250-pF, 5-kV Ceramic Transmitting Capacitor	3¾" Turns ¼" Copper Tubing 2¼" Inside Diameter, 2" Long Tap Approximately 1¼ Turns from Ground End

Note: Adjust L1 tap using directional coupler or reflectometer (see text).

All three transformers are identical in construction and must be tuned to the same resonant frequency. Table 7-3 gives the values for C1 and L1 for operation on the 15- and 20-meter bands.

To match a 50- or 75-ohm coaxial line to the transformer, connect the transmission line from the transmitter to the input terminal of the directional coupler or the reflectometer as shown in Fig. 7-10. Use a short (12 inches or so) length of test cable from the directional coupler or reflectometer output to the antenna coil. The short test cable should be the same impedance as the transmission line and fitted with a PL-259 plug at one end and a copper alligator clip at the other. The outer conductor of the test cable and the bottom of the antenna coil should be connected to ground (radial system) through a short, heavy, copper-braid strap or other low-inductance conductor. Feed sufficient rf signal into the main transmission line *at the transmitter end* and adjust the tap connection to the coil for lowest line swr. The tap connection will be very critical, but with careful adjustment the line swr can be reduced to 1.1:1, or better, at the resonant frequency of the LC circuit. The swr will rise at frequencies removed from the resonant frequency. With the high-C transformers shown, the swr is still less than 2:1 at 50 kHz either side of

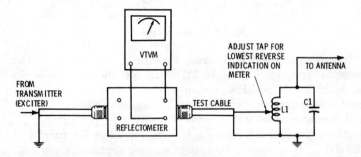

Fig. 7-10. Using the reflectometer to adjust tap on tuned rf autotransformer.

resonance. With low-C transformers, there will be a much less rapid rise in the swr at frequencies removed from resonance. The main reason the author uses the high-C transformers is that he is primarily interested in DX work. When receiving, the sensitivity of the system is slightly better at the resonant frequency of the tuned circuit. The 14-MHz phasing conditions and radiation pattern for the three-element vertical array are shown in Fig. 7-11.

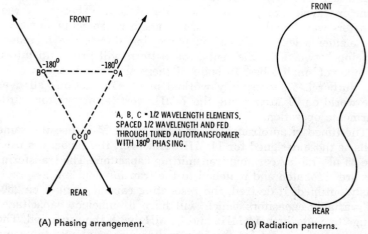

A, B, C = 1/2 WAVELENGTH ELEMENTS, SPACED 1/2 WAVELENGTH AND FED THROUGH TUNED AUTOTRANSFORMER WITH 180° PHASING.

(A) Phasing arrangement.　　　　(B) Radiation patterns.

Fig. 7-11. Phasing arrangement and approximate radiation patterns for 40-meter three-element vertical array operated at 14 MHz.

For DX work, arrays with half-wavelength elements have several definite advantages over those using quarter-wavelength elements. In the first place, the current loop (maximum) for the half-wavelength element is half way up the element toward the top. On the higher-frequency bands this is important because maximum radiation occurs at a point high above the ground, reducing ground losses in the vicinity of the array. The base of the half-wavelength element is a relatively high-impedance point compared to the base of a quarter-wavelength element and the array does not depend upon an *image* antenna in the earth for *balance*. Therefore, the ground and radial requirements for the half-wavelength-element array are somewhat less stringent than for the quarter-wavelength-element array. The higher-impedance feedpoint requires less rf current and the feedpoint resistance losses are correspondingly reduced. The circulating rf current in the tuned autotransformer, however, is fairly high and the circuit should be of low-loss construction with the coil mounted in the clear.

While 40-meter three-element arrays with quarter-wavelength elements can be operated on the 21-MHz band, some problems are encountered. The quarter-wavelength elements at 40 meters become *three-quarter* wavelengths at 21 MHz and the spacing between the elements also increases to three-quarter wavelengths. Each element will have *two* current loops, one will be located one-third of the way down from the top and the other will be at the base or feedpoint. The ground or radial-system requirements are stringent if ground losses are to be kept low. The quarter-wavelength phasing lines at 7 MHz now become three-quarters wavelength at 21 MHz. In some arrangements, however, the elements will be driven *in phase,* but due to the increased spacing between the elements, the pattern will break up into a number of small lobes. In spite of these shortcomings, however, the author has successfully worked many DX stations all over the world on 21 MHz using the 7-MHz vertical arrays for third-harmonic operation.

The tuned rf autotransformers used for 21 MHz use the same coils as those designed for 14 MHz. However, the capacitors used are 25-pF, 7.5-kV ceramic transmitting capacitors. The transformers are resonated and matched to the transmission line as previously outlined. If desired, the resonating capacitors may be 250-pF variable capacitors which will have a sufficient capacitance range to cover both 14 MHz and 21 MHz with the same coil. The tap connection point for the 50-ohm line was at almost the same place on the coil for both bands. A "happy-medium" point can be found where the swr will be low on both bands and it will only be necessary to readjust the resonating capacitor when changing

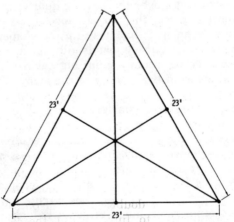

Fig. 7-12. Layout for three-element 21-MHz array using half-wavelength vertical elements with half-wavelength spacing between elements.

bands. The advantage of using fixed-capacitance resonating is that the transformers may be permanently tuned for each band and then switched by a remote-control relay.

The layout for a three-element array using half-wave elements and half-wave spacing for the 21-MHz band is shown in Fig. 7-12. At the author's station, two elements are driven in phase with each other and the third element is driven at 180° out of phase with respect to the other two elements. The pattern is a broad figure 8 and the radiation angle in the vertical plane appears to be very low. The use of this array has been limited because of poor propagation conditions on 21 MHz at the time this book was written, but all indications are that it will be very effective for DX work. The matching system uses tuned rf transformers as described in the foregoing discussion.

7-5. A FIVE-WATT SIGNAL GENERATOR FOR THE VECTORSCOPE

As mentioned previously, the low-level signal generator is perfectly satisfactory for use with rf bridges and similar sensitive devices. However, the quadrascope (vectorscope) requires approximately 100 volts peak-to-peak of rf voltage for full deflection of the beam. The modern ssb exciter usually has a TUNE or CARRIER INSERTION position on the mode selector and can supply a steady unmodulated signal. The Collins S-line exciter used by the author can be adjusted to a suitable level for making phasing adjustments with the quadrascope. However, the author prefers to use a separate rf signal source.

The schematic for the five-watt signal generator is shown in Fig. 7-13. Oscillator stage V1 is crystal-controlled and produces test signals on the 80-, 40-, 20-, 15-, and 10-meter bands. A 3.550-MHz crystal is used to produce rf output at 14.200 and 21.300 MHz. A 3.600-MHz crystal produces rf output at 7.200 and 28.800 MHz. The 40-meter test signal is taken from the oscillator plate circuit by means of a two-turn link winding over the "cold end" of coil L1. The rf output on the 20-meter band is obtained from the plate circuit of V2. The link coil has two turns coupled to the "cold end" of coil L2. In this case, V2 operates as a frequency doubler. To obtain rf output on the 15-meter band, V2 is operated as an amplifier with its plate circuit tuned to 7.100 MHz. Output stage V3 then operates as a tripler with its plate circuit tuned to 21.3 MHz. A 10-meter test signal is obtained by operating both V2 and V3 as frequency doublers. The rf output on all bands is more than sufficient to fully deflect the 902A crt in the quadrascope. If rf output is desired on the 80-meter band, the plate circuit of V1 may be tuned to the crystal frequency.

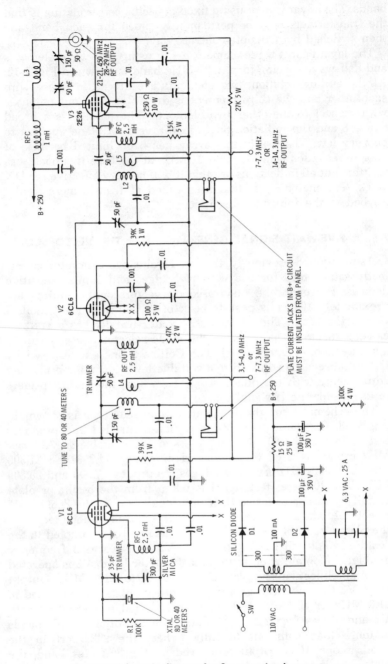

Fig. 7-13. Schematic diagram for five-watt signal generator.

Obviously, the 5-watt signal generator could be used as a low-power cw transmitter if the plate supply voltage was higher. With a B+ of 250 volts on the plate of V2 and 450 to 500 volts on the plate of V3, the rf power output from V3 should be in the order of 25 to 30 watts. However, no provisions for keying or the elimination of key click have been included in the circuit shown here. Refer to one of the radio-amateur handbooks for appropriate keying and click-filter circuits.

Other types of tubes will operate in this circuit. The types shown just happened to be in the author's junkbox.

Coaxial Cable Data

FORMULAS

1. Characteristic Impedance Z_o

$$Z_o = \frac{138}{\sqrt{K}} \; \log_{10} \; \frac{D}{d}$$

where,

Z_o is the characteristic or surge impedance of the line,

D is the inside diameter of the outer conductor,

d is the outside diameter of the inner conductor expressed in the same units as D,

K is the dielectric constant of the insulating material.

2. To Convert Length in Degrees to Length in Feet, Type-A Dielectric

For 360° $L = \dfrac{649.44}{f}$ For 135° $L = \dfrac{243.54}{f}$

For 270° $L = \dfrac{487.08}{f}$ For 90° $L = \dfrac{162.36}{f}$

For 180° $L = \dfrac{324.72}{f}$ For 45° $L = \dfrac{81.18}{f}$

where,

f is the frequency in MHz.,

L is the length in feet.

3. To Convert Length in Degrees to Length in Feet, Foam Dielectric

For $360°$ $L = \dfrac{777.36}{f}$ For $135°$ $L = \dfrac{291.51}{f}$

For $270°$ $L = \dfrac{583.02}{f}$ For $90°$ $L = \dfrac{194.34}{f}$

For $180°$ $L = \dfrac{388.68}{f}$ For $45°$ $L = \dfrac{97.17}{f}$

where,
  f is the frequency in MHz,
  L is the length in feet.

4. Example: Find the length in feet of an RG-8/U type-A dielectric coaxial cable 90° long at 7.2 MHz.

$$L = \frac{162.36}{7.2}$$

$$= 22.55 \text{ feet}$$

PHASING CABLE LENGTHS

Table A-1. Phasing Cable Lengths in Feet for Amateur-Band Center Frequencies 3 to 30 MHz (Standard A Dielectric)

Degrees	3.75 MHz	7.15 MHz	14.175 MHz	21.225 MHz	28.6 MHz
360°	173.18	90.83	45.82	30.55	22.70
270°	129.89	68.12	34.36	22.95	17.03
180°	86.59	45.42	22.90	15.30	11.35
90°	43.30	22.71	11.45	7.65	5.68
45°	21.65	11.36	5.73	3.83	2.84
22.5°	10.83	5.68	2.87	1.92	1.42
10°	4.81	2.52	1.27	0.85	0.63
5°	2.41	1.26	0.64	0.43	0.32

FLEXIBLE COAXIAL LINES

Table A-2. Specifications for Flexible Coaxial Lines

Type	Impedance (Ohms)	Dielectric Type	Velocity Constant	Maximum RMS Voltage (Volts)	Maximum RMS Current (Amperes)
RG-8/U	52	A	0.66	4000	7 to 10
RG-11/U	75	A	0.66	4000	5 to 7
RG-58/U	53	A	0.66	1900	2.8 to 5
RG-59/U	75	A	0.66	2300	2.8 to 5
RG-17/U	52	A	0.66	11,000	12 to 15
RG-62A/U	93	A	0.84	750	— — —
RG-71B/U	93	A	0.84	750	— — —

COAXIAL CONNECTORS

Table A-3. Standard UHF Single-Contact Coaxial Connectors

Type	Military Number	Amphenol Number	Description	For Cable Type RG-
PL-259	PL-259	83-1SP*	Cable Plug	8, 11
UG-203U	UG-203U	83-776	Cable Plug	58
UG-111U	UG-111U	83-750	Cable Plug	59, 62, 71
UG-176U	UG-176U	83-168	0.257" I.D. Reducer	59, 62, 71
UG-175U	UG-175U	83-185	0.207" I. D. Reducer	58
SO-239	SO-239	83-1R†	Panel Receptacle	
UG-106U	UG-106U	83-1H*	0.345" I.D. Hood	8, 11
UG-177U	UG-177U	83-765*	0.155" I.D. Hood	58
UG-646U	UG-646U	83-1AP	Right Angle	
PL-258	PL-258	83-1J†	Double Female Adaptor	
UG-363	UG-363	83-1F	Feed Thru	
M-358	M-358	83-1T	Tee	

* silver-plated, all others Astroplate finish
† female connector, mates with all plugs in UHF series

Air-Wound Coils

Sometimes it is difficult to find a Miniductor coil with the exact inductance value needed for some specific purpose. A good example would be a multisection filter where a 11.5-microhenry inductance must fit in the space of a 1-inch cube. The coil diameter must be less than an inch for clearance. Barker & Williamson Miniductors provide a means of obtaining such coils. They are designed to be cut to provide the inductance needed and are available in various sizes.

HOW TO CUT A MINIDUCTOR

To obtain the coil needed in the previous example, refer to Table B-1 to select the proper inductor. In this case, there are four possible choices: types 3004, 3008, 3044, and 3012. A little rough mental arithmetic lets us eliminate 3004—it's obviously too long. A quick calculation eliminates 3008—11.5 microhenries will be about 64 percent of 18 microhenries, and it would take more than half of the 2-inch length making it too long.

Using the same procedure for 3044 shows us that 1 inch of the 3044 Miniductor has the needed inductance but leaves nothing for mounting and may touch the ends of the 1-inch cube. Further calculation for 3012 shows that 29 percent of the 3-inch length gives us 0.87 inch. To find the number of turns required, take 29 percent of the total number of turns in the full Miniductor which equals .29 × 96, or approximately 28 turns. Allow two additional turns for mounting and connecting leads (one turn at each end). You should try 30 turns of Miniductor 3012. The graph in

Fig. B-1 gives a quick approximation of the percentage of inductance of a shortened length of Miniductor compared to the inductance of a standard length.

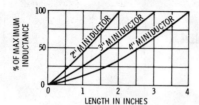

Fig. B-1. Proportional inductance of Miniductors versus length.

Table B-1. Specifications for Barker & Williamson Miniductors

Type No.	Coil Diameter (Inches)	Turns per Inch	Coil Length (Inches)	Wire Size	Inductance (Microhenries)
3001	½	4	2	18	0.18
3036	½	6	2	18	0.40
3002	½	8	2	18	0.72
3037	½	10	2	18	1.10
3003	½	16	2	20	3.00
3038	½	24	2	22	6.75
3004	½	32	2	24	12.00
3005	⅝	4	2	16	0.275
3039	⅝	6	2	18	0.620
3006	⅝	8	2	18	1.100
3040	⅝	10	2	18	1.700
3007	⅝	16	2	20	4.500
3041	⅝	24	2	22	10.000
3008	⅝	32	2	24	18.000
3009	¾	4	3	16	0.620
3042	¾	6	3	18	1.400
3010	¾	8	3	18	2.500
3043	¾	10	3	18	3.900
3011	¾	16	3	20	10.000
3044	¾	24	3	22	23.000
3012	¾	32	3	24	40.000
3013	1.0	4	3	16	1.000
3045	1.0	6	3	18	2.400
3014	1.0	8	3	18	4.100
3046	1.0	10	3	18	6.600
3015	1.0	16	3	20	17.000
3047	1.0	24	3	22	38.000
3016	1.0	32	3	24	68.000
3017	1¼	4	4	14	2.300
3048	1¼	6	4	14	5.000
3018	1¼	8	4	16	9.000
3049	1¼	10	4	18	14.000
3019	1¼	16	4	18	36.000
3050	1¼	24	4	22	81.000
3020	1¼	32	4	24	145.000

Table B-1 (continued)

Type No.	Coil Diameter (Inches)	Turns per Inch	Coil Length (Inches)	Wire Size	Inductance (Microhenries)
3051	1½	4	4	14	3.100
3052	1½	6	4	14	7.000
3053	1½	8	4	16	12.500
3054	1½	10	4	18	20.000
3055	1½	16	4	20	50.500
3056	1½	24	4	22	110.000
3057	1½	32	10	24	200.000
3021	1¾	4	4	14	4.200
3058	1¾	6	4	14	9.400
3022	1¾	8	4	14	16.500
3059	1¾	10	4	16	26.000
3023	1¾	16	4	18	67.000
3060	1¾	22	4	22	150.000
3024	1¾	24	4	24	270.000
3061	2	4	10	12	15.000
3025	2	6	10	12	33.000
3026	2	8	10	14	59.000
3027	2	10	10	16	92.000
3062	2	16	10	16	238.000
3063	2½	4	10	12	22.500
3029	2½	6	10	12	51.000
3030	2½	8	10	14	90.000
3031	2½	10	10	16	140.000
3064	3	4	10	12	32.000
3033	3	6	10	12	71.000
3034	3	8	10	14	125.000
3035	3	10	10	16	198.000

APPENDIX C

Manufacturers and Suppliers

CERAMIC CAPACITORS

High Energy Corp.
Malvern Industrial Park
Malvern, Pennsylvania 19355

Request latest catalog and name and address of nearest distributor

RELAYS

Potter and Brumfield Division
AMF Incorporated
1200 East Broadway
Princeton, Indiana 47670

Request latest catalog and name and address of nearest distributor

OTHER ITEMS

Burstein-Applebee
3199 Mercier Street
Kansas City, Missouri 64111

Mail order and retail parts house. All kinds of electronic equipment.

Lafayette Radio Electronics Corp.
111 Jericho Turnpike
Syosset, New York, 11791

Mail order and retail parts house. All kinds of electronic equipment.

Barker & Williamson, Inc.
Canal Street
Bristol, Pennsylvania 19007

Manufacturer of amateur radio equipment, dummy loads, power meters, etc.

Melvin L. Heineken, K5MNJ
Box 353
Edgewood, New Mexico 87015

Custom printed wiring boards. Special custom test instruments.

Bell Industries
J. W. Miller Division
19070 Reyes Avenue
Compton, California 90221

Components manufacturer. Request catalog and name and address of nearest distributor.

Amidon Associates
12033 Otsego Street
North Hollywood, California 91607

Iron powder and ferrite cores. Write for data sheets and prices.

R. L. Drake Company
540 Richard Street
Miamisburg, Ohio 45342

Amateur radio products.

Palomar Engineers
Box 455
Escondido, California 92025

Amateur radio antenna products, cores, and matching transformers, etc.

NOTE: The inclusion of the above manufacturers and suppliers and their mailing addresses is merely for the convenience of the reader. This listing *does not* constitute any endorsement of either the manufacturer, supplier, or his product. However, the author has purchased supplies from all of the above with satisfactory business relations.

APPENDIX D

Answers to Self-Examinations

Chapter Two
1. False
2. False
3. True
4. False
5. False
6. True
7. True
8. True
9. False
10. True

Chapter Three
1. False
2. True
3. True
4. True
5. True
6. False
7. True
8. False
9. True
10. False

Chapter Four
1. True
2. True
3. False
4. True
5. True
6. True
7. True
8. True
9. False
10. True

Chapter Five
1. True
2. False
3. False
4. False
5. True
6. True
7. True
8. True
9. True
10. True

Chapter Six
1. False
2. True
3. True
4. True
5. True
6. True
7. True
8. True
9. True
10. True

Index